SpringerBriefs in Molecular Science

SpringerBriefs in Molecular Science present concise summaries of cutting-edge research and practical applications across a wide spectrum of fields centered around chemistry. Featuring compact volumes of 50 to 125 pages, the series covers a range of content from professional to academic. Typical topics might include:

- A timely report of state-of-the-art analytical techniques
- A bridge between new research results, as published in journal articles, and a contextual literature review
- A snapshot of a hot or emerging topic
- An in-depth case study
- A presentation of core concepts that students must understand in order to make independent contributions

Briefs allow authors to present their ideas and readers to absorb them with minimal time investment. Briefs will be published as part of Springer's eBook collection, with millions of users worldwide. In addition, Briefs will be available for individual print and electronic purchase. Briefs are characterized by fast, global electronic dissemination, standard publishing contracts, easy-to-use manuscript preparation and formatting guidelines, and expedited production schedules. Both solicited and unsolicited manuscripts are considered for publication in this series.

Guillaume Graciani

Effective Communication and Single-Molecule Research

Springer

Guillaume Graciani
SAMOVAR, Télécom SudParis
Institut Polytechnique de Paris
Palaiseau, France

ISSN 2191-5407 ISSN 2191-5415 (electronic)
SpringerBriefs in Molecular Science
ISBN 978-3-032-10425-0 ISBN 978-3-032-10426-7 (eBook)
https://doi.org/10.1007/978-3-032-10426-7

This Springer imprint is published by the registered company Springer Nature Switzerland AG
The registered company address is: Gewerbestrasse 11, 6330 Cham, Switzerland

Pour Loann et Swann.

Preface

This book is largely concerned with the idea of communicating science effectively. In a time when public trust in science faces unprecedented challenges, communication not only determines how the work of scientists is received, but whether it can fulfill its potential to improve human understanding and well-being.

This manuscript is a direct continuation of a simple 60-second video, which was recognized by Nature's Science in Shorts award, a prize that highlights the power of clear science communication and the growing need for a true representation of who scientists really are, diverse in their backgrounds and aspirations, similar in their biases and way of navigating the world, through the lens of the scientific method.

Inspired by that experience, I set out to reflect more broadly on what it means to communicate science effectively: not merely simplifying, but making ideas visible, relatable, and truthful, without losing their essence or sacrificing accuracy or nuance.

Alongside practical considerations and reflections on communication strategies, I explore scientific examples from my own work, explaining how science is actually done in a lab today, in the field of single-molecule research.

Palaiseau, France
August 2025

Guillaume Graciani

Competing Interests The author has no conflicts of interest to declare that are relevant to the content of this book. The author is employed by the Institut Mines Telecom and works at SAMOVAR, Télécom SudParis, Institut Polytechnique de Paris, 91120 Palaiseau, France.

Contents

Part I
The Need for Science Communication

For me, it is far better to grasp the Universe as it really is than to persist in delusion, however satisfying and reassuring.

—Carl Sagan, The Demon-Haunted World: Science as a Candle in the Dark

Chapter 1
Introduction

Abstract Science education is inseparable from the personal voice of the educator, shaped by lived experience, cultural background, and their sense of empathy. This diversity is not optional, it is essential. It reminds us that science belongs to everyone, and that learning actually benefits from a plurality of approaches. Still, there are common grounds to reach a global audience, and whatever the medium, the clarity and humility required to teach well remain the same: no simplifications that betray meaning, no condescension, no shortcuts. The Covid-19 crisis starkly revealed the cost of poor communication. Despite public interest, many essential scientific concepts remained misunderstood. Yet, models of transparent communication in some parts of the world showed that effective education was possible when explanations were not treated as mere afterthoughts but as essential for the implementation of public policies. This chapter argues for a renewed commitment from scientists to speak clearly and directly, not to simplify science, but to reveal its depth through grounded, accessible explanations.

1.1 TL;DR

Science lives in the space between discovery and understanding. Not just the understanding of those who make science, but the understanding of everyone whose lives are touched by scientific progress.

Throughout these pages, we will explore why science communication has become not just important, but vital to the future of scientific enterprise itself. In an era where public trust in science faces unprecedented challenges, our ability to communicate effectively determines not only how the work of scientists is received, but whether it can fulfill its potential to improve human understanding and well-being.

This manuscript is both a guide for fellow educators seeking to share their work more effectively, and a bridge for anyone interested in understanding how scientific knowledge moves from the bench to the broader world. After all, science communication is ultimately about recognizing that discovery means little if it remains isolated from the communities it could serve.

G. Graciani, *Effective Communication and Single-Molecule Research*,
SpringerBriefs in Molecular Science,
https://doi.org/10.1007/978-3-032-10426-7_1

This book unfolds in two parts. The first reflects on why science communication matters now more than ever. The second sheds light on how science is actually done up close in a research lab, through the lens of single-molecule research and biophotonics.

In Part One, we begin with my personal experience. From the early online tutoring of classmates to being a physics professor, we reflect on the pedagogical landscape that is available to both learners and teachers: how diverse educators offer diverse paths to understanding, and why that variety is essential, not only for inclusion, but for what it adds to the act of learning itself. We explore how empathy in teaching creates the conditions for genuine understanding, and how every explanation, no matter the audience, deserves precision and care.

The COVID-19 pandemic becomes a case study for being a missed opportunity to educate, even as the public's curiosity surged. Concepts like exponential growth, conditional probability, or vaccine trials entered everyday conversation, but often without the clarity they needed. In contrast, examples like South Korea's daily public scientific briefings showed that when scientists speak directly, people listen and understand.

The following chapters argue that the stakes of science communication are not merely academic. In a world saturated with misinformation, where urgent issues like artificial intelligence, climate change, and energy demand informed decision-making, the scientist's voice must rise in volume and in clarity. For the public to be able to take part in open science, make informed decisions, create new policies, and understand the technological world in which it lives, it must learn the scientific method. Yet communication is not merely a service to society, it also benefits science itself. Science is a collective enterprise that needs to be shared to fully reach its potential. By sharing their work, scientists also define which facets of science should enter in the public conversation and, ultimately, where funding should be directed and what policies should be implemented.

We then examine strategies for effective communication. Analogies, stories, visual intuition; not as simplifications, but as acts of translation. We learn to anchor abstractions in the familiar, to show not just what science knows, but how it comes to know it through science history and the personal motivations of the scientists behind the scene who made it. Above all, we learn to respect our audience: to be exacting without being exclusionary, and clear without being reductive.

Part Two turns inward, toward the laboratory. Here, we move from the need to communicate science to the nature of the science being done. Research, we are reminded, is a profoundly human endeavor. Scientists are not immune to bias or error, but they are bound by an ethic of honest inquiry. We demystify the process of publication, of peer review, and of reading scientific literature, a system that should be open and taught as much as possible to the public.

As a practical example, we then focus on a powerful frontier: the probing of single molecules with light. Using biophotonics techniques such as magnetic tweezers, cavity amplified scattering spectroscopy, and mass photometry, we explore how light can reveal the most delicate interactions between proteins, and how such tools help uncover the origins of diseases or point toward new treatments. Real-world examples, from studying membrane protein folding to accelerating antibody drug conjugate development, demonstrate the practical impact of this field.

Finally, we look ahead. Emerging technologies, new measurement platforms, and deeper biological insights all suggest that the age of single-molecule science is only beginning. The closing reflections tie everything together: how the scientist, as both communicator and explorer, can help reshape the public's relationship with knowledge by inviting others to see the world through their lens.

In this book, the story of science is not only about discovery, it is about the deep responsibility to share that discovery, honestly, and patiently, with anyone who is willing to listen.

1.2 Speaking from Experience

Every educator speaks with a distinct voice shaped by their sensibility, their cultural backgrounds, and their life paths as both learner and teacher of their own. This diversity is not incidental, it is essential for two reasons. First, it reminds us that science is not the domain of a particular archetype, but rather that scientists are not cast from a single mold, they are members of our society, as diverse as they could be. Second, the multiplicity of teaching voices offers a profound gift to learners: the chance to encounter knowledge through perspectives that best suit their needs and ways of learning. Whether through conferences, books, short or long format videos, there is a trend forming that each student can find their own learning approach, going as far as even using large language models as tailor-made, personal tutors.

Every teacher, too, carries their own sense of what it means to truly convey knowledge. For my part, I have come to believe that empathy, and the ability to put oneself in someone else's shoes are crucial. To teach well is to imagine, without judgment, the questions forming in another person's mind. The cost of making any judgment, is to potentially stop their momentum toward understanding. Even the smallest assumption or step in the wrong direction may become a barrier to understanding. In the end, learning is most powerful when done in solitude, on one's own terms, in quiet reflection. Learners should be given enough space to reflect and form ideas in their own mind, not prefabricated, but constructed on their own. As Peter O'Toole once put it, learning is a journey taken "*alone, with no interlocutor, an unobserved, uninhibited private study.*"

I had the chance to understand that very early on. Around the age of twelve, I began tutoring fellow classmates, not in the classroom, but online, through the spartan live text chats of the early internet. Back then, there was no room for ornament. Explanations had to be clear, direct, concise, limited by time-restricted paid internet and the patience of typing. Having virtually exactly the same background and experience as other students made me realize how understanding the way your audience thinks was crucially important in order to find the right teaching words.

Explanations also had to be free of condescension, with no space for evasive metaphors and convoluted explanations, or the lazy comfort of simplifications that trades accuracy for ease. Educators cannot afford being condescending: every explanation or answer to questions requires the same involvement and precision, even through countless repetitions. Also, effective explanation does not mean dilution. It means precision, patience, and no omission of details. One does not dumb down science but reveals it carefully, re-framing the unknown in the light of the known, with explanations grounded in reality. Done well, this act of translating scientific jargon into relatable experience does not obscure complexity, it illuminates it.

That experience, humble and informal, became the foundation of how I later approached science communication including, years later, Nature's Science in Shorts competition. The goal of that contest was to make a single-minute video from scratch explaining an aspect of one's research to the general public. Every part of it, the script, the pace of the video and editing, what was shown on screen and composition, as well as the music used were all designed to convey the essence of a message, in a clear and straightforward manner. The contest was designed to reach as wide as an audience as possible, but also to entice all sorts of researchers to take part in it. Showing how research is actually conducted in labs today, and how researchers actually look like in all of their diversity.

It is unfortunately not surprising how such a contest came to exist today and why the motivations behind it were sound. As an example and from a purely pedagogical standpoint, the Covid-19 crisis revealed more than a global health emergency. In many parts of the world, it became at best a missed opportunity for education, at worst a pedagogical failure, in a moment when the need for public understanding was urgent, and yet, too often, the explanations fell short. For once, the questions people asked were not abstract. They were immediate, deeply real. People genuinely wanted to understand what incidence rates meant. They were shown curves on national television, and in many cases, they watched misread explanations on what an exponential growth is. It wasn't indifference to science, but rather confusion, and it was avoidable. It was also the time to explain who scientists really are, and what kind of work they produce: how science is conducted, reviewed, published.

We have seen, at the time, a remarkable collective effort by the scientific community. Researchers across disciplines mobilized, not only to find answers, but to explain them. Suddenly, the inner workings of science were no longer confined to academic circles: the process through which science is made, corrected, and published was in the open. Terms like "peer review," "blind study," "clinical trial," entered the public lexicon. And yet, even then, explanations on how vaccines are developed and tested remained mostly elusive. Another example is the concept of conditional probability, and how the result of a test can be interpreted was rarely communicated with the clarity it deserved. A test with a high accuracy does not mean that a positive result guarantees infection. But a second positive test? Then the odds shift, and Bayes' theorem begins to speak. These were teachable moments, that unfortunately were left on the side.

Contrast this with what occurred in some other parts of the world, such as South Korea, where scientists themselves addressed the nation each day on television, explaining openly what was known, what remained uncertain, and what was expected. The result was a more cohesive public response, more easily accepted by people because they understood it, not because the science was different, but because the trust in it was earned, slowly and transparently.

Throughout this book, we will return to this idea: that scientists must speak louder in volume, and more clearly, more often, and more directly. Not to proclaim certainties, but to share the rigor of the process by which we come to understand.

Chapter 2
Science Needs a Voice

Abstract In an era of unprecedented access to information, the internet has opened gateways to scientific knowledge on a global scale. Yet alongside this accessibility comes a rise in misinformation, cognitive fatigue, and the erosion of public trust in science. This chapter explores the paradox we live in: a time when truth and falsehood coexist in the same digital spaces, indistinguishable without critical thinking. It examines the challenges of distinguishing scientific research, rooted in doubt and revision, from established scientific knowledge, which deserves trust and confidence. Beyond facts, this chapter argues for teaching scientific thinking: the mindset of curiosity, honesty, and methodical inquiry that empowers individuals to navigate this complexity. Drawing on personal experiences, educational studies, and cognitive psychology, it highlights the barriers many face, not of intellect, but of confidence, and the phenomenon of learned helplessness that keeps people from engaging with science. This chapter advocates for greater public access to primary scientific literature, supported by clearer writing and better education on how to read scientific texts. Ultimately, it calls on scientists not only to inform, but to bridge the gap between knowledge and society and to reawaken the public's capacity to learn and question.

2.1 Now More Than Ever

The total sum of human knowledge, once bound in monasteries or academic archives, is now instantly accessible to billions. A curious mind can summon the mechanics of the Brownian motion, the mathematics of quantum entanglement, or the structure of a black hole with a few keystrokes. From encyclopedic platforms like Wikipedia to research articles increasingly available through open-access initiatives, the once-exclusive corpus of scientific literature has, in theory, been placed into the hands of the global citizen.

More than this, we have entered the age of democratized expression: voices proliferate on every platform, transmitting knowledge through long-form documentary series to thirty-second social media posts. Machine learning algorithms, chat-bots, and advanced natural language models now start to offer synthetic overviews of

G. Graciani, *Effective Communication and Single-Molecule Research*,
SpringerBriefs in Molecular Science,
https://doi.org/10.1007/978-3-032-10426-7_2

scholarly knowledge in a way that keeps on getting better. For the attentive and discriminating observer, this is a golden age of intellectual availability.

And yet, we are surrounded, perhaps more than ever, by confusion. The open frontier of the internet, like all frontiers, is not only fertile but wild. It teems not just with truth but with deliberate misdirection. It has never been easier to fall into rabbit holes of misinformation, algorithmic feedback loops that reinforce untruth, sometimes subtly, sometimes aggressively. What begins as curiosity can quickly become entrapment.

This distortion is not always the product of ignorance. It can be intentional. Misinformation campaigns, sometimes automated through bot networks, sometimes organic and viral, seed falsehood with precision. At the same time, the sheer volume of available information makes it difficult to evaluate claims. Fewer and fewer people take the time—because they simply don't have it—to verify or cross-reference what they read. There is a growing fatigue, a saturation of content that encourages passive reception over critical engagement. Add to this the monetization of authority: digital certification, once a marker of verification, can now be purchased, casting further doubt on the reliability of sources.

In such a landscape, memory itself becomes fragile. Events are misremembered, or reinterpreted by those with vested interests. It is not uncommon to see individuals rewrite the past in a manner that flatters their current stance, secure in the knowledge that few will recall the precise facts, and fewer still will take the time to check. The result is a kind of epistemological drift, where perception overrides reality, and narrative supplants evidence.

Against this backdrop, an unsettling question arises: why do certain scientific ideas, like Newton's laws of motion, achieve near-universal acceptance, while others like anthropogenic climate change are met with skepticism or outright denial? Why are some domains of science granted the status of objective truth, while others are treated as matters of opinion?

The answer may lie, in part, in a fundamental misunderstanding of science itself. Scientific research is a process defined by uncertainty, by questioning, by doubt. It is a method of systematic inquiry, where even the most beautiful theories must survive confrontation with evidence. But scientific knowledge—those results which have been repeatedly tested, refined, and confirmed—deserves a different kind of trust. When people confuse the two, when they treat all scientific statements as provisional, or all findings as up for debate, they fail to distinguish between the living edge of research and the solid trail of understanding and knowledge it leaves behind.

This, I believe, is where scientists must step forward. We have a responsibility to act as bridges between the turbulent, iterative nature of research, and the stable clarity of established science. In schools across the world, science education often halts somewhere in the nineteenth century. Students learn of Ohm and Newton, of Boyle and Curie, but modern science, and the ideas that shape today's research, rarely enter the classroom. As a result, students leave school with only little foundation for understanding contemporary scientific discourse.

This is not solely a failing of the educational system. It is also a failure on the part of the scientific community. We have not done enough to explain not only what we are doing in our laboratories, but why it matters, and how it connects to the lives of those outside our institutions. Without such efforts, the gap between the public and the scientific enterprise widens.

Consider France, where the international TIMSS study has revealed, year after year, a sobering trend: scientific literacy is declining, and the response from both political leaders and the general public has been marked by indifference. The consequence is not merely academic, it threatens the very fabric of informed citizenship in a world increasingly shaped by scientific decisions and complex technologies.

And yet, there is hope. Clear, thoughtful science communication can influence public policy, reshape educational priorities, and fuel innovation. But it must go beyond the superficial. Today, as we confront the profound consequences of technologies like machine learning and artificial intelligence, we are also confronting questions of agency. What does it mean to delegate judgment to machines? What are the ethical implications of automation in decision-making? These are not questions that belong only to computer scientists. They are questions for everyone.

Climate change, too, demands a shared vocabulary. In countless conversations, I have heard sincere citizens speak of "producing" or "consuming" energy in ways that betray the fundamental concept of conservation. Yet those misunderstandings are not the fault of the public. They are symptoms of a deeper issue: a society where scientific terms have become disconnected from their meanings.

We must reconnect them with clarity and purpose. The promise of science lies not only in discovery, but in dialogue. And it is through that dialogue that we may yet illuminate the path ahead, a dialogue that increasingly becomes a necessity for the society we live in.

2.2 Success, Failure, and Bias

> It is so shocking to find out how many people do not believe that they can learn, and how many more believe learning to be difficult.—Frank Herbert, Dune

How does science shape society? How far does it reach into the fabric of our culture, our politics and policies, our decisions, and our daily lives? In truth, the diffusion of scientific thought into the broader world is a process not easily measured. When dealing with science communication though, the most important metric might be when someone manages, somehow, to spark interest in science in a single individual.

We often hear of people whose path toward science began not in a laboratory, but in a story. A book, a podcast, a documentary viewed late at night: these seemingly modest moments become catalysts. They awaken a sense of wonder, a desire to understand, and above all, a belief that such understanding is possible. Science communication, when done with sincerity and clarity, can spark vocations. It can inspire. And not only science journalism or popular essays: even science fiction can achieve this. By projecting the present into possible futures, it gives form to curiosity.

And yet, we must acknowledge a limitation that plagues even the best-intentioned outreach. We are often preaching to the choir. Those who actively seek out scientific content, whether videos, interviews, or essays, are often those already drawn to its music. They arrive curious, already halfway to engagement. In this, science communication sometimes becomes a mirror, reflecting existing interest rather than generating new connections.

This reveals a deeper structural challenge, one not of content, but of inclusion. A significant portion of the population feels estranged from science. They do not believe it belongs to them. This sense of exclusion is often seeded early, in school. A poorly taught lesson, a discouraging grade, an impatient teacher: such moments take root and cast long shadows. And the wound is not only cognitive, it is emotional and it lingers, becoming a belief that "science is not for me."

Writer Frank Herbert once observed something profoundly resonant: "It is so shocking to find out how many people do not believe they can learn, and how many more believe learning to be difficult." This is a quiet tragedy of modern education. Not only in science, but across all domains of learning: literature, music, philosophy, chess, or any thoughtful pursuit. There persists the illusion that some minds are simply not equipped to learn. But learning is not an inborn gift reserved for the few. It is a process, a discipline, a decision.

I remember my own experience as a student in France's classes préparatoires, where we were tasked with mastering a vast and rigorous body of mathematics. We memorized entire proofs, absorbed definitions, theorems, corollaries and lemmas across sprawling topics, and were trained in both speed and precision. Much of the technical content has faded with time. But what remains, what still guides me as a scientist today, is the foundational confidence that I can learn. I can approach the unfamiliar without fear. When I encounter a concept outside my domain—for instance, a new experimental tool I could use for my research—I do not hesitate. I begin, and I learn. It is this mindset, more than any one fact, that education must instill.

But for many, the opposite mindset takes hold. Learned helplessness, a concept developed by Martin Seligman and Steven Maier in the1960s describes a psychological condition in which repeated exposure to failure, especially when it seems outside one's control, leads to resignation. It becomes internalized: "I cannot succeed, no matter what I do." This belief then persists even when the opportunity for success becomes real.

There is a striking classroom experiment that illustrates this: two groups of students are presented with a series of anagram puzzles. Unbeknownst to them, the first two puzzles given to Group A are impossible to solve, while Group B receives solvable ones. When both groups then receive a common, solvable word, students in Group A often fail to find the answer. Discouraged by early failure, their confidence collapses. They no longer even attempt. Meanwhile, Group B, whose early experiences affirmed their ability, succeed.

It is a simple demonstration, but one that echoes a much broader issue. We must recognize that confidence in learning is not distributed equally. It is shaped by experience, but also by privilege. Some children grow up in rich and encouraging environments, where parents can explain homework, offer examples, suggest books. Others are not given these tools. And so, when the moment of learning arrives, they doubt themselves before the explanation even begins.

What, then, can we do? I believe the solution lies in more direct access to science, not just through interpretation or translation, but through the real materials of research. The scientific article, the primary source, should not be seen as a fortress closed to all but experts. It should be a gateway. But this requires two parallel efforts.

On the one hand, scientists must write with accessibility in mind. This does not mean sacrificing precision. But it does mean reducing unnecessary jargon and remembering that clarity serves both reader and writer. Why not, for example, systematically including guiding introductions for the general public in all published papers? On the other hand, we must teach people how to read scientific literature. Because it is a genre unto itself, highly structured, filled with conventions, written in a rhythm that is unfamiliar to most.

There is, for instance, a method to approaching a scientific paper. One does not begin on page one and march forward. Rather, you start with the title, to locate the domain. Then the abstract, to see the broader aim. If the topic still intrigues, and if the reader is not a novice, then the figures, carefully studied, offer a visual synopsis. The conclusion next, to grasp the scope and limits. Only then, if one seeks depth, the full reading. For the unfamiliar concept, the first bibliographic reference in the introduction is a beacon, an invitation to follow the thread backwards, until understanding is built.

Today's tools make this even more feasible. Large language models can produce informed summaries tailored to any level of expertise. With proper use, they become allies in reading, helping bridge the gap between complexity and comprehension. But we must also demystify the broader process of publication. The public should know how scientific articles are written, how authors collaborate, how work is submitted, peer-reviewed, revised, published. They should understand that articles are funded, often publicly, and that this information is disclosed within the text. They should learn how journals are ranked, how access is controlled, how knowledge flows, and where it is dammed.

This book will return to these questions. But the premise is simple: science is not the property of the academic world. It is a method, a history, a shared inheritance. If people are to engage with it meaningfully, they must be invited not only to listen, but to read. To question. To learn believing, perhaps for the first time, that they can.

Chapter 3
Scientists Have Vocal Cords

Abstract This chapter reflects on the recent fate of the Palais de la Découverte, a unique Parisian institution that embodies democratic access to science. More than a museum, it offers direct, rigorous encounters between the public and practicing scientists, spaces for dialogue, curiosity, and shared understanding. Its renovation aimed at deepening this mission, only to be abruptly halted, signaling a troubling retreat from science as a shared civic resource. This possible loss is emblematic of broader challenges: declining scientific literacy, public disengagement, and institutional shifts away from foundational education in favor of trend-driven themes. Drawing on André Gorz and Kohei Saito's distinction between "open" and "locking" technologies, this chapter warns against the alienation of the general public from increasingly complex technological solutions being implemented into our society, if not scientifically well-educated. Without robust science communication, these technologies risk deepening democratic deficits. The chapter argues for reclaiming open, participatory spaces where citizens are trusted with complex scientific topics, invited into scientific discourse, and given tools to understand and shape the technologies that define their lives. At stake is more than pedagogy or policy, it is the principle that science must remain a shared endeavor, embedded in public life, and open to all who seek to question and learn.

3.1 The Democratization of Science

In the heart of Paris stands a place devoted not to art, or politics, or commerce, but to science. The Palais de la Découverte is not a museum in the conventional sense, though it exhibits wonders. Nor is it merely a lecture room, though learning echoes through its corridors. It is something far rarer: a living space for science to meet the public, not through spectacle nor conventional teaching, but through genuine encounter.

Unlike many cultural institutions which treat science as a historical artifact, the Palais embraces science as a process: ongoing, evolving, unfinished. Its rooms do not merely house exhibits, they hold conversations. With amphitheaters in which scientists speak directly to curious visitors, lectures where complex ideas are not

G. Graciani, *Effective Communication and Single-Molecule Research*,
SpringerBriefs in Molecular Science,
https://doi.org/10.1007/978-3-032-10426-7_3

diluted but made accessible, and demonstrations where experiments are performed in real time, by real researchers, with all the rigor of the scientific method. But more than any of these features, it is the spirit of proximity that makes the Palais unique: the real encounter with science, up close.

In recent years, this institution had embarked on an ambitious renovation, not simply architectural, but philosophical. The goal was to deepen its accessibility, to rethink how science is shared and who feels welcome in it. Scenographers collaborated with researchers, designing new spaces where dialogue could thrive. These were not just exhibition halls but spaces of exchange: rooms built for discussion, for public debate, for the simple but powerful act of asking questions and receiving thoughtful answers.

The project's heart was clear: to foster a culture in which science is not an external authority but a shared inquiry. It was a rare alignment of vision and practice, of public engagement and scientific honesty. And then suddenly it stopped.

As of August 2025, this vision, though near realization, has been suspended. The government, without clear justification, removed the president overseeing this institution. There is now serious talks of redirecting the place toward new themes: artificial intelligence, or the intersection of science and art. Both are worthy subjects: rich, timely, essential in their own ways. But in privileging these over the foundational work of making fundamental science accessible to the public, we risk losing something irreplaceable.

This is not just an administrative decision. It unfortunately very much looks like a symptom. It reflects a deeper uncertainty, perhaps even an unease, about the place of science in French society today. As previously discussed, France has long been slipping in international measures of science literacy, with the TIMSS study revealing year after year a troubling decline. Yet there is little public outrage, little political response, and the silence is telling.

The fate of the Palais de la Découverte is thus emblematic. In a moment when clear, honest, and inclusive scientific dialogue is urgently needed, not only in classrooms but in the public square, an institution that embodied those values is being reshaped into something else, something less focused on democratic access to knowledge. What is lost is not only a building, or a program, but an idea: that science, in all its rigor and difficulty, belongs to everyone.

This leads us to a broader philosophical concern, one articulated with particular clarity by the Austrian-French philosopher André Gorz. In his critique of productivism, the ideology that values growth and output above all, Gorz introduced a distinction that particularly resonates here: the difference between open and locking technologies.

Open technologies are those that foster exchange, transparency, and shared agency. They enable participation. They encourage learning. They bring people together in the service of a common understanding. Locking technologies, on the other hand, divide. They are closed by design, monopolized by expertise, shielded by security or proprietary knowledge. Their logic is vertical: they concentrate control in the hands of the few, rendering the many passive.

This framework has been revisited by Kohei Saito in his recent manifesto on degrowth [1], where he applied it to today's most pressing technological debates. Take, for instance, nuclear energy. While it is a powerful and low-emission energy source, it is a quintessential locking technology. It requires centralized management, strict secrecy, and high technical specialization. Decisions are made top-down. The public has little insight, and even less input.

In a society where science communication is robust and where citizens are taught not only the facts of science but how science is practiced, such technologies could be subject to meaningful democratic oversight. But in the absence of that foundation, they are governed in isolation. The result is not only a lack of transparency but a quiet disempowerment. Citizens feel alienated from decisions that affect their lives, and science becomes something done to them, rather than with them. It forbids the percolation of science into society.

Saito extends this concern to climate change technologies, particularly those that promise to remove carbon from the atmosphere—so-called Negative Emission Technologies—and to geoengineering proposals aimed at manipulating planetary systems. While these innovations may have technical merit, they often embody the logic of locking. They are conceived and implemented in elite circles, dependent on specialized knowledge, disconnected from democratic deliberation.

Even more troubling, as Saito warns, is that such technologies may displace more sustainable solutions. They create the illusion of control, of technical salvation, while crowding out the possibility of a different future, one based not on managing the symptoms of crisis but on transforming the systems that cause it. In his words: "Technology ends up suppressing and pushing out the possibility of creating a completely different mode of living in the face of the impending crisis, of forming a society truly independent of fossil fuels."

Here, again, the role of science communication is vital. Not as a tool for persuasion, but as a framework for shared understanding. For the public to meaningfully engage with questions of nuclear policy, artificial intelligence, or climate change, they must first be invited into the language and logic of science itself. They must be trusted with complexity. They must be given not only information, but the means to inquire, to have access to the processes, to the debates, to the uncertainty that defines real inquiry.

This is what the Palais de la Découverte once offered, and what we risk forgetting. When we strip away spaces where science is not only seen but experienced, when we sideline institutions that bring the public and researchers into shared dialogue, we do more than damage a building: we erode the possibility of a democratic science.

To move forward, we must reclaim that possibility. We must insist that science be not only rigorous, but participatory. That its doors remain open. That the next generation of citizens, not just of scientists, encounters science not as a mystery locked away, but as an invitation. As something they have a right to question, to understand, and to shape.

Because in the end, the future we choose will depend not only on the technologies we invent, but on the conversations we dare to have about them.

3.2 The Benefits of Science Communication

We have seen that, on many levels, communication is not just an option, it is a responsibility for researchers. But there is more to it; it is also directly useful to scientists themselves, in the context of how science is currently made.

In science mediation, we often speak of discoveries, of theories tested and knowledge refined, of the patient and careful unraveling of nature's deepest secrets. But behind every equation, every model, every dataset, there is a voice. A human voice that is curious, uncertain, hopeful, trying to make sense of the world. And that voice, if it is not heard, vanishes into the void.

Communication, then, is not an embellishment to the scientific endeavor. It is not a hobby to be indulged once the "real" work is done. It is, in truth, an essential part of the scientific process itself. We have spoken earlier of the societal necessity of science communication: how the public, in a world steeped in complexity and misinformation, needs interpreters of the scientific method, not to preach truths, but to share the process by which we come to understand. But there is a second, more personal axis to this responsibility: communication is a tool, powerful and practical, for the scientist as well.

Since I began engaging more actively in science communication: publishing videos, speaking in interviews, participating in podcasts, broadcasting my research results, the effects have been unmistakable. New opportunities began to open. Not theoretical possibilities, but concrete collaborations, invitations, professional intersections that might never have occurred otherwise. Work in industry, consulting roles, and even unexpected creative collaborations with artists. Not because I sought them directly, but because my voice had reached places where my publications alone had not.

More subtly, visibility has brought connections within the research world itself. It has facilitated collaborations, initiated conversations across disciplines, and, perhaps most importantly, made it easier to recruit students. For a researcher, the ability to attract curious, motivated minds is not a small matter. Interns, Ph.D. candidates, postdoctoral fellows: they are the lifeblood of a laboratory. And in a world where the volume of research is overwhelming, being known can be the deciding factor that brings the right student to your door.

Let us not confuse this visibility with vanity though. It is not about self-promotion in the superficial sense. It is about the recognition that science, though grounded in rigor and objectivity, is still conducted by humans in a network of trust, dialogue, and awareness. To be known in your community as a reviewer, a contributor, someone whose work resonates, is to be embedded in that network in a way that sustains careers and opens doors.

This is especially true for early-career researchers. During the precarious years of postdoctoral work, when the future is uncertain and competition fierce, a well-articulated voice can make a difference. It will not replace good science, nor should it, but it can distinguish you in a sea of equally competent peers. Later, as a professor, it becomes a means not only of attracting collaborators and students, but also of gaining access to funding. Visibility, paired with credibility, builds trust in your ability to lead projects, manage resources, and contribute to the broader goals of scientific institutions.

But there is a larger structural issue at play: the sheer volume of scientific output. Each year, millions of papers are published across thousands of journals. The deluge is unrelenting, and the competition for attention: funding, citations, media interest, is intense. In such an environment, it is not enough to quietly produce excellent work. One must also make the case for why that work matters. Not in a grandiose or exaggerated way, but with clarity, precision, and a willingness to engage. Otherwise, your findings risk being lost in the noise.

This is not without its challenges. It requires effort (sometimes considerable effort) and precious time to translate research into forms that can be shared widely and effectively. And yet, we must not fall into the false dichotomy that depth and communication are mutually exclusive. They are not. It is possible to conduct serious, sustained research and still speak about it in a way that is intelligible beyond a narrow circle. It is a balance, but one that is well worth cultivating.

Another axis of balance lies in the question of trends. Scientific research is not conducted in a vacuum. It is shaped by culture, by politics, by funding priorities, and by the shifting attention of the public. Some topics rise to prominence, drawing resources and interest, others fade into the background. As researchers, we must navigate this landscape with care.

On the one hand, there is great value in following one's intellectual passions. The most profound discoveries often come not from chasing trends, but from sustained attention to questions that are personally compelling. Professor Steven Strogatz has put it beautifully: work on what you love, and what is central to you will become central to your work. This is not a sentimental notion. It is a practical insight. Passion sustains focus. It keeps the mind turning over a problem long after the grant proposal is submitted. It invites serendipity, allows for patience, and creates the conditions under which deep understanding can flourish.

The case of Alain Aspect, Nobel laureate in physics, illustrates this principle. His groundbreaking work on quantum entanglement did not happen overnight. It took nearly a decade of meticulous experimentation, years of slow progress, doubt, recalibration. What enabled this was not a sudden alignment with a funding trend, but a lab environment that gave him the freedom and time to follow the path of curiosity. "The freedom and the means that were given to me at the Institut d'Optique," he later said, "were the keys to my success." Such freedom is rare, and growing rarer still.

For many researchers around the world, especially in institutions driven by short-term metrics and constant evaluations, there is immense pressure to conform to trending topics. To work on whatever is currently attracting funding, media interest, or citation potential. This pressure is not always explicit, but it is palpable. It shapes decisions, subtly but powerfully, and often at the expense of long-term, curiosity-driven inquiry.

I have seen laboratories where even tenured professors are under relentless pressure to publish in high-impact journals, to apply for prestigious funding, to stay visible. The result is often a narrowing of ambition: researchers become adept at chasing metrics, but less able to pursue the slow, strange questions that may yield tomorrow's breakthroughs. It is a system optimized for productivity, but not necessarily for discovery.

And yet, paradoxically, it is also public engagement that shapes the trends. National policy, funding bodies, institutional priorities: all of these are influenced, in part, by what society deems valuable. And what society deems valuable is, in turn, shaped by what it understands and what it hears. When scientists speak clearly and compellingly about their work, they do more than inform the public. They shape the future of their own disciplines.

In that sense, science communication is not only about translation. It is about participation. It is how researchers engage not just with lay audiences, but with the very structures that support or hinder their work. It is a way of taking part in the collective decisions about what questions we ask, what problems we prioritize and fund, and what futures we imagine.

Of course, this engagement must be tempered with humility. No single scientist can predict the future. No researcher can know, in advance, which ideas will bear fruit. But in speaking openly about their process, their uncertainties, and their hopes, scientists contribute to a culture in which knowledge is not hoarded, but shared, where the pursuit of understanding is seen not as a solitary climb, but a common journey. In that journey, communication is not a detour but a path itself.

3.3 Life Navigator

The trajectory of society hinges not merely on the accumulation of knowledge, but mostly on the widespread cultivation of a particular way of thinking. We have spoken already of the necessity of teaching science, of making it accessible, of restoring trust between researchers and the society that funds, follows, and is shaped by their work. This is no small task. As we face the intertwined urgencies of climate change, the global energy transition, and the rapidly encroaching frontiers of artificial intelligence, the need for broad scientific understanding has never been more pressing. But I would like to take the argument one step further. It is not just scientific facts we must teach, it is scientific thinking.

There is a distinction here that is subtle, but profound. It is one thing to understand at time t, say, the concept of CO_2 accumulation in the atmosphere, but it is quite another to develop the habit that pertains to the scientific method and to incorporate it into one's life. The courage to doubt. The discipline to test. The humility to admit error. The impulse to revise one's opinion and understanding in the face of new evidence. These are not facts to be taught, but practices to be cultivated. They form a posture toward the world. And that posture, if broadly adopted, has the potential to reshape our public life.

In my own experience, it was not any single scientific concept that changed how I navigate the world. It was the process of learning itself. Looking back, I cannot claim to recall every detail of my science curriculum or all the obstacles overcame during my Ph.D. But what remains is more important: the training of attention, the perseverance in the face of complexity, and the quiet confidence that I can, when necessary, explore and learn again. That knowledge, once out of reach, can be drawn closer with time and care. This confidence is not unique to science, of course. It applies equally in any domain where mastery emerges slowly, with effort and repetition. But science, in particular, formalizes it. It gives us a structure, a collective agreement, on how to pursue truth while knowing we may never reach it fully.

To learn the scientific method, then, is to learn a way of living. It is to accept that certainty is rare, that knowledge is provisional, that every hypothesis is an invitation to be wrong in a more interesting way. And yet, paradoxically, it is also a source of extraordinary clarity. For in science, even doubt has discipline. Even uncertainty has structure.

The philosopher Karl Popper once described science as the "friendly-hostile cooperation of citizens of the knowledge community." A curious phrase, but an apt one. Science is not a solitary endeavor. It is a social enterprise. It is argument, counterargument, and change of mind. It is claim and replication. It is correction, refinement, and, sometimes, transformation. And its only demand, the one non-negotiable, is honesty. Not perfection, not omniscience, not even neutrality, but honesty when facing doubt.

Etienne Klein, the French philosopher of science, has expressed this simply. The scientist's task, he says, is to be sincerely interested in a question, to seek answers, to produce results, and to compare those results with others interested in the same question. There is no need to transcend the human condition. As scientists, we bring our subjectivities, our biases, our fallibilities into the laboratory. But what we must also bring is our integrity. We must mean what we say. And we must allow our conclusions to be tested by others.

This firm and persistent ethic sits at the heart of science. It is also, unfortunately, what makes science so ill-suited to the current machinery of public discourse.

We live in an age of infotainment, of soundbites and manufactured outrage, when society tends to favor absolutes and certainties. Our talk shows are not spaces of dialogue, but of clash: every expert has their own opinion, impossibly certain, and nowhere it is to be found a guest that is measured or moderate. Certainty is rewarded, nuance is punished. A guest who admits "I don't know" is unlikely to be invited back. A thinker who expresses doubt is seen not as rigorous, but as weak. On social media, everyone is so sure about every aspects of their lives. Doubt is very hard to come across.

This creates an impossible dilemma for the scientific voice. For doubt is not a weakness in research, it is its motor. It is what drives inquiry. But in a media environment addicted to spectacle, doubt is drowned out. There is little room for the patient explanation, the hesitant hypothesis, the refusal to speculate beyond what the data allows. One is expected to take a position, and to defend it vigorously, even theatrically. And so the scientist, who trades in probabilities and error bars, who measures rather than proclaims, is often left out of the conversation altogether.

We saw this vividly during the COVID-19 crisis. The public craved clarity, understandably so. But research, by its very nature, could not offer certainty. It could offer evolving models, best estimates, and evidence-based recommendations. But these were not always welcomed. In the competition between the loud and the careful, the careful often lost. And yet, it was the careful who were right.

This is not an argument for scientists to become entertainers. But it is a reminder that we must find new ways to speak. If we succeed, the rewards are enormous. A public trained not only in scientific facts, but in scientific thinking, becomes harder to manipulate. It becomes more resilient against misinformation. It becomes better equipped to participate in democratic decision-making on issues that are, at their core, technical: climate, energy, technology, health.

These are not skills reserved for scientists. They are tools for every citizen. And if we can teach them well, if we can model them in our public life, we may find that we are not only better equipped to face our challenges, but better able to imagine what lies beyond them. But now a question remains: how to teach the scientific method? The answer is by shedding light on the people behind the discoveries, their motivations, their process. A hint of a solution is detailed in the next chapter.

Reference

1. Saito, K.: Marx in the Anthropocene: towards the idea of degrowth communism. Cambridge University Press (2023)

Chapter 4
Strategies for Communicating Science

Abstract In today's digital age, widespread access to science communication may fuel curiosity, but it also fosters a subtle form of forgetfulness, or "digital amnesia," where audiences remember *where* they learned something, but not exactly *what*. Inspired by Betsy Sparrow's research on cognitive offloading, this chapter explores how the internet has become a vast external memory, reshaping how we learn and retain scientific ideas. As creators and educators, we must recognize this effect: fleeting engagement with polished content, like YouTube videos, rarely yields deep understanding unless reinforced through active teaching and structured reflection. To counter this drift, this chapter proposes strategies to teach the scientific method over scientific facts. A method grounded in context and personal motivations. We begin with "why," anchoring every concept in its historical origin, human story, and real-world motivations. We tailor messaging to specific audiences, using precise analogies, while carefully marking their limits. We structure our presentations into digestible segments: one idea, image, slide, or sentence at a time, and anticipate and answer questions before they become distractions. We aim to cultivate learner confidence through early successes, avoiding jargon traps and over-simplifications that breed distrust. Ultimately, meaningful science communication demands humility, patience, and a commitment to reveal not just what we know, but why we ask the questions, and how inquiry unfolds. By doing so, we transform passive consumption into active curiosity and lasting understanding.

4.1 Digital Amnesia in Science Communication

Can science communication actually be useful at all to teach scientific concepts? We have already covered the fact that there are biases: mostly people who are already attracted to learn about science actively seek scientific communication-related content. Still, among this population, is science communication an effective form of conveying and retaining information?

G. Graciani, *Effective Communication and Single-Molecule Research*,
SpringerBriefs in Molecular Science,
https://doi.org/10.1007/978-3-032-10426-7_4

In the grand unfolding of human knowledge, the printing press once shattered the monopoly of memory. Now from books, to librairies, and to the internet, each step has placed our collective intelligence even more within reach, while subtly rewriting the architecture of thought itself. In 2011, psychologist Betsy Sparrow and her colleagues set out to examine how the omnipresence of information, accessible in an instant, might be reshaping the very function of memory [1]. They introduced what is called "digital amnesia", a concept I, like many of us, have experienced personally.

Their inquiry was elegant and deceptively simple. Participants were asked to absorb trivia. But there was a twist: some were told the facts would be saved, stored, retrievable later. Others were told the facts would vanish. The results were startling: when the human mind believes knowledge is safe elsewhere, it declines the invitation to store it within. This relates to two concepts: *transactive memory*, being less likely to remember the content of information if one believes it will be available later, and *cognitive offloading* (especially true with the internet), when people are treating external systems as an extension of their memory.

And yet, this is not forgetfulness born out of laziness but quite the opposite. When the facts were saved, subjects became especially adept at recalling *where* they could find them, much like we remember which physics book holds an important equation, or which friend knows the name of that elusive progressive metal band we once saw live. The study revealed that our minds, confronted with boundless digital access, are adapting. We are not losing memory but we are relocating it.

This phenomenon, now dubbed the Google Effect, strongly echoes the concept of transactive memory: a shared cognitive system where individuals remember who knows what. In the digital era, that "who" is no longer a person, it is a website, a forum, a video online, an invisible extension of ourselves.

Coming back to science communication, when dealing with intricate and complex concepts, this form of amnesia is especially true. I, for one, will remember precisely if I saw a YouTube video on a certain subject, but not the details of the content itself if I haven't studied it seriously or, better, taught it to my students. The issue: I tend not to retain information since I know it is easily accessible and, if needed, I can always go back to it to find answers. This is easily understandable when you have experience as a teacher: when addressing a class, one has to constantly explain the same concepts over and over again, using different explanations, analogies, images, exercises for the students to retain information.

In my opinion and in that sense, science communication having the goal of teaching science is better when allowing people to learn the scientific method, rather than teaching the scientific facts themselves. This is especially true when facts are so easily accessible. Eventually, the core content of a complex theory will be lost in translation, and only a vague souvenir of important notions will remain.

In that context, there are some strategies that educators can use to better transmit how to think scientifically. An invaluable tool in the misinformation age.

4.2 Strategies for Educators

4.2.1 Distraction as Friction

When we speak of science communication, we are not merely talking about the transfer of information, we are talking about making a connection: bridging a mind to a concept through carefully chosen words, vivid metaphors, and finely made visuals. Science communication seeks to render the abstract meaningful, to translate data into understanding, and to awaken curiosity where once there was only confusion.

There is no better example of this than the Science in Shorts contest organized by Nature. A contest that strips communication down to its essentials. In 60 seconds, you have to spark curiosity, introduce a subject, and explain why it matters and how it is done. When I entered that contest, I approached it as an experiment in distilled communication. The script was divided into bite-sized components, each a single thought, carried by a single sentence. Each segment began with a question or image designed to pique interest, followed by a brief expansion: "This is what I study. This is why it matters. This is how I do it." Each phrase was a standalone seed, nurtured just enough to make contact, but not so much as to spread tangential roots.

The visual structure followed the same logic. Each idea was paired with a distinct framing: a shift in angle, a zoom into my expression, a diagram fading in and out so that the audience's attention could refresh, reboot, and refocus. The camera moved with intent, never staying static long enough to allow distraction. The entire video was an exercise in unity of form and function: each cut, each sentence, each image served a single, transparent purpose: to convey one idea with minimal friction, and the most potent friction of all is distraction.

I knew that attention is not simply what is held but rather, it is what is freed. If you leave questions unanswered in people's minds: "Why are we even doing this?" "Why is this method chosen?", they will dwell on those questions instead of listening. Therefore it is important to add meaning to everything that is said. Indeed, the students, or the audience at large, need to know *why* we do things. The whole point, to me, is that if those questions are not addressed, the audience will keep on having them in their mind, wondering why things are done in the certain way that is displayed and, ultimately, this will distract them from what is being said. Like a presentation slide that contains too many things, forcing the audience to read in their mind and not to listen to a word you are saying, you should try your absolute best to free their mind by answering all the questions that may linger so that they can turn their full, undisturbed attention to your explanation. So every sentence in my video had to anticipate the next question, and every visual had to anticipate the next doubt. It became, ultimately, a dance: guide the attention in, answer the questions before they arise, and then move deeper, all within sixty seconds.

That unity of idea, script, and visual, may be extreme in a contest video. But it holds a lesson for all forms of communication: a talk, an article, a classroom session. We owe our listeners more than information. In order to reduce friction to learning, we must clear the mind for new ideas by addressing the most likely sources of distraction. Coming back to Chap. 1, this requires teachers to place themselves in their student's shoes and a sense of empathy. We must show, we must explain why we show, and we must do so knowing our audience.

4.2.2 Know and Trust Your Audience

Scientific communication is not a monolith. The person reading a grant proposal differs from the viewer of a YouTube video, and both differ from the high school student hearing your talk. To speak effectively is to acknowledge that difference and to tailor your message accordingly.

I learned this early on as I presented the same concepts to people from different backgrounds: art students, engineers, physics graduate students. I never used the same slide deck twice. Because in every room there is a different prior knowledge, a different perspective, and a different vocabulary for curiosity. The core message might be the same, but it must live in forms that resonate.

For instance: explaining the Shannon-Nyquist sampling theorem to cinema or photography students requires a different entry point than to physics majors. To photographers, I ask: "Have you ever tried to film a jump-rope routine in stop motion? If you only capture frames when the rope is either up or down, the result looks frozen. You need at least twice that amount of frames if you want to convey any sort of motion." That principle is the same as signal sampling: there is a lower limit to the number of samples you should take from an analog signal if you want to recreate it faithfully. The analogy is simple and familiar to them, yet it is absolutely correct. Once the analogy is used efficiently to defuse the seemingly difficult and confusing concept, students are then more inclined to jump into the real subject. First of course with simple maths functions like sines, then with more elaborate things such as images and videos.

This is, however, not a license to oversimplify or to sell the audience an illusion of understanding that later unravels as confusion or betrayal. I remember, from early middle school math classes, how simplifications were presented as fact, then abandoned a year later to introduce other concepts. It sowed distrust. Worse: when students later encounter a fuller explanation, they feel deceived, as though the initial teaching was a lie. We must trust our audience to grow in understanding. We must neither underestimate their intelligence nor overwhelm them with unfamiliar jargon.

Oversimplification, not only bringing defiance, can bring fundamental misunderstanding that can be very hard to fight and re-established. Etienne Klein often brings up the misunderstanding surrounding Heisenberg's uncertainty principle. Taught in schools—at least in France—it is stated that quantum mechanics basically demonstrates that we cannot know with an arbitrary precision, both the speed and position of a particle. Reading that, most people believe that a particle has a speed and a position well determined, but that, somehow, quantum mechanics prevents us from measuring it accurately. However, what Heisenberg meant, was that quantum particles are never small objects moving through space. One cannot measure those quantities because they simply do not possess them. This fundamental misunderstanding is firmly anchored in people's mind and, I for once can see it first hand when I am teaching quantum mechanics to students. I believe it would not have been much more difficult to explain the details of the indetermination principle, as it was intended by Heisenberg.

4.3 Adding a Sense of Meaning

The single most important element in commanding the attention of students is avoiding distraction by answering the question "why?". Fundamentally, we lean toward stories that make sense of the world. When we introduce a theorem, reveal a law, or demonstrate a phenomenon, students want to know: why are we introducing this theorem right now? Why were the researchers even interested in that subject to start with? In short, students need to have a sense of background and purpose.

When we pull a result "out of a hat," we risk making the subject feel arbitrary, disconnected, or unmotivated. Students grow frustrated, sensing that they are shoehorned into mathematical or scientific constructs without a context or a reason. But when we take the time to unfold the *why*, the historical impetus, the human curiosity, the pressing questions of the era, then mathematics and science become less like rigid instruments and more like living conversations stretching across time.

I am a big believer that the history of science should be taught alongside science itself, because the classroom should not feel like a museum of frozen results, but rather like the unfolding of human inquiry. The best historical movies bring characters to life by immersing us in their personal motivations; so too should science classes animate the figures behind the ideas. Take quantum mechanics, for instance. Its genesis lies in what was then called the "ultraviolet catastrophe." Classical physics, when pushed to the limit, predicted that a blackbody would emit infinite energy, an absurd result. To resolve this paradox, Max Planck introduced the idea that energy might be quantized. He did not pluck that from thin air but addressed a real crisis in the scientific world. And from there, blossomed the quantum century. This is the kind of narrative that electrifies learning.

How did Einstein and Bohr wrote letters to each other and discussed through scientific manuscripts about their respective interpretations of quantum mechanics? How John Bell came into play, what he thought of this debate and how he contributed to it? What was the goal of Alain Aspect and how did he design his experiment to test John Bell's equations that got him a Nobel prize? This is just a tiny example on one very specific aspect of physics, but literally every single aspect of science is filled with such anecdotes.

We should start teaching science through the scientists who made it. Not only does it add empathy, a better sense of identification and of who they are as people, it also tremendously help understand underlying motivations about the "why" of science. Even more, it teaches through concrete examples how science is conducted, what the scientific method is. And here, modern media: videos, podcasts, video games, can play a decisive and complementary role. Popular channels like Veritasium are increasingly embedding their explanations within narratives: the context, the culture, and the personal histories behind scientific discoveries. This enriches understanding without sacrificing rigor.

In the same vein, I advocate teaching physics, chemistry, and maths all together in a coherent, single natural science program. Explaining how the discovery of Newton's laws of gravity are closely linked to his invention of calculus and so on and so forth. For a story to be meaningful, it needs to be coherent. Very often, when tutoring high school students, have I amazed them by making connections between physics and chemistry phenomena, when those concepts were initially taught to them as completely separated topics.

Of course, this cannot be done in a single form of teaching, because there is actually a big corpus to cover, and that is why we need to consider the importance of other media in teaching science in a complementary way.

This brings us to a deeper truth: science is not a static body of knowledge. It is a process. A journey. A craft practiced in laboratories, debated in labs and journals, revised through critique and replication. That journey is what Part II of this book is all about.

Reference

1. Sparrow, B., Liu, J., Wegner, D.M.: Google effects on memory: cognitive consequences of having information at our fingertips. Science **333**(6043), 776–778 (2011)

Part II
How Science is Actually Done in the Lab Today

If you wish to make an apple pie from scratch, you must first invent the universe.
—Carl Sagan

Chapter 5
Inside the Science Lab

Abstract What does science look like from the inside? This chapter offers a guided tour through the daily realities of scientific research, from the inner workings of a laboratory to the intricate journey of publishing a paper. It begins with the structural foundations: how scientists, driven by curiosity but constrained by resources, compete for funding by proposing precise, impactful research plans. Once funded, labs operate as collaborative ecosystems where experiments are designed, refined, and often reimagined over months or years at a time before yielding meaningful results. The chapter then explores the critical role of scientific publishing, not just as a metric of productivity, but as a public act of transparency. From pre-prints to peer review, from funding declarations to data sharing, each stage reinforces science's collective pursuit of truth. Yet this process is often opaque to outsiders. By demystifying the structure of scientific articles, abstracts, figures, metadata, and supplementary materials, this chapter also equips readers to engage critically with scientific literature. In doing so, it emphasizes that science is not a static body of knowledge, but a living, iterative conversation. One where transparency, trust, and participation are vital. To understand science is not merely to learn facts, but to see how knowledge is made and shared.

5.1 Where Scientific Ideas Are Born

Scientists are no exception to the human condition: they stumble, they presume, they are biased, and have cultural backgrounds. But despite these very human limitations, there emerges a striking commonality: a deep and abiding honesty. Not just honesty in the moral sense, but an epistemic integrity, an honesty of inquiry. They are people genuinely interested in a question, compelled not merely by ambition or obligation, but by an insatiable desire to understand. And they do not act alone.

Science thrives in community. It is not content to rest in the mind of the solitary thinker, its findings must be tested, examined, confronted by others equally engaged. When we speak of the scientific method, we often reduce it to a sterile checklist: hypothesis, experiment, analysis, and conclusion. But beneath this framework lies a great human story, a story of cooperation, disagreement, refinement, and consensus,

G. Graciani, *Effective Communication and Single-Molecule Research*, SpringerBriefs in Molecular Science,
https://doi.org/10.1007/978-3-032-10426-7_5

carried out not in isolation, but in the shared spaces of laboratories, scientific journals, and conference halls. So we might now ask: how is this grand collaboration, this ongoing conversation about nature, actually organized at the level of the modern scientific laboratory?

Structures vary from nation to nation, but a common framework emerges. Most scientists working in academia begin their journeys not as independent investigators, but as students and apprentices, eventually rising, if chance and perseverance allow, to the role of professor or principal investigator. In such a role, the scientist is no longer merely a learner of knowledge but becomes a guide, a leader, a mentor, and a project manager. Their position is typically supported by their host institution, a university or research center, which provides a salary, access to a laboratory, and perhaps a modest internal budget. But to do research of real consequence, to ask bold questions and chase subtle answers, requires something more: funding.

And here we arrive at one of the more intricate, sometimes fraught, elements of modern scientific life. Funding is the lifeblood of research. But it is not distributed arbitrarily. It is, almost without exception, competitive. Scientists must apply for funding from a variety of sources, some governmental, some philanthropic, some internal to universities, others offered by national or international research councils. These funds are not simply granted on the basis of good intentions. The process requires the scientist to articulate a coherent vision: who are they, and what have they done before that qualifies them to take on this challenge as a scientist and a project manager? What, precisely, is the question they seek to answer, and how will they go about investigating it? What resources will they need, and why? How will the funds be used: on personnel, on equipment, on travel, on publication fees? And most importantly: what benefit might this research bring to science, to society, to humanity?

A research proposal, then, is both a blueprint and a testament. It asks the scientist to show their thinking not only to a panel of peers but to the future: to demonstrate that this inquiry is not merely plausible, but promising. Should the proposal be accepted, the funds are allocated but even then, oversight continues. Not all expenses are permitted, and detailed reports must be submitted periodically. The scientific imagination must now coexist with administrative vigilance.

With funding in place, the real work begins. Professors assemble a small team composed largely of graduate students and postdoctoral researchers. While collaboration between labs and across institutions is very common, much of the day-to-day progress, the careful calibrations, the long hours of coding or data collection or sample preparation, is carried out by these junior scientists, under the careful supervision and mentoring of their senior colleagues.

In experimental sciences, the rhythms of the lab are steady and persistent. Each week, data is gathered, painstakingly, often with uncertainty. Team meetings serve as a forum where preliminary results are presented, dissected, and discussed. A single research project can take years from conception to fruition. First comes an insight, a hypothesis or an intuition. Then follows the construction of an experimental design, the building or repurposing of instruments, the initial collection of data. These early efforts are rarely definitive. Noise creeps in. Designs must be revised. Controls must be added. What at first seemed like a clear signal may turn out to be an artifact of some overlooked variable.

Science, then, is not a straight road to truth. It is a winding path, full of detours, wrong turns, and necessary delays. But eventually, with persistence and cooperation, something emerges: data that resists dismissal, patterns that hold up under scrutiny. At that point, the team turns its attention to communication: to writing the story of their findings in the form of a scientific manuscript.

Here we reach yet another vital juncture in the scientific process: publication. The act of publishing in science is not peripheral, it is central. It is not merely a way of showcasing success, it is the mechanism by which science verifies itself. Through publication, scientists offer their findings to the world, not as final truth, but as claims open to challenge and refinement.

Of course, in today's academic climate, the imperative to publish also has a pragmatic dimension. Careers are built, and sometimes stalled, on publication records. The number of articles, the prestige of the journals, the citation counts: all these are used, rightly or wrongly, as proxies for scientific impact. Decisions about hiring, tenure, and future funding often hinge on such metrics. It is a system that aims to reward productivity, but it can sometimes distort incentives.

Still, beyond the metrics and career pressures, there remains a deeper reason to publish: to share. As the musician Charlie Fink once put it, "What you share with the world is what it keeps of you." In that spirit, scientific publication is a gift to the collective human endeavor. Results do not exist in the world if not shared.

Yet for all its idealism, the publication process is complex and sometimes opaque. Consider the modern trajectory of a scientific manuscript. Once a study has been completed and the findings written up, the authors often post a version on an open-access pre-print server such as arXiv. This version, though publicly available, has not yet undergone peer review. It is a kind of public draft, a way of announcing to the community that a result has been achieved and inviting early feedback.

Simultaneously, the authors submit the manuscript to a scientific journal. Each journal serves a different audience and subject area, so the selection of the journal is itself strategic. Alongside the manuscript, the authors include a cover letter to the editor: a brief but vital document that summarizes the significance of the work and explains why it is a good fit for the journal. In a sense, the story of their research.

If the editor deems the submission worthy of review, the paper is sent anonymously to a set of peer reviewers, usually three experts in the field. These reviewers act as the guardians of scientific quality. Their role is not merely to praise or criticize, but to interrogate. Are the methods sound? Are the conclusions supported by the data? Are there alternative explanations that have not been considered? The reviewers may suggest revisions, from minor clarifications to major overhauls. Sometimes they recommend rejection.

If the reviewers call for revision, the authors must respond, often by conducting new experiments or reanalyzing their data. This process can repeat more than once and last for months. It can be exhausting, even discouraging. But when it works as intended, peer review improves the quality of the final publication. It filters out error and exaggeration. It forces scientists to justify their claims.

Once accepted, the revised manuscript becomes the "version of record." This final version is often made available in repositories like HAL, an open-access archive that ensures the findings remain freely accessible. Yet it's worth noting that this version may differ significantly from the initial pre-print. Months of revision, clarification, and editorial feedback may have reshaped the argument, refined the figures, even altered the conclusions.

The final act in the publication process is formatting and dissemination. The journal prepares the article for release: correcting grammar, ensuring figures clarity, adjusting layout. Some journals make the article freely available (if the authors or their institutions can pay a publication fee sometimes worth thousands of dollars). Others place the article behind a paywall, accessible only to those with subscriptions.

This system, imperfect, slow, and sometimes costly, nonetheless constitutes one of humanity's greatest achievements. It is a global mechanism for accumulating reliable knowledge. It is a collective memory, continually revised and expanded. It is how we know what we know, and how we remember what we have discovered.

In this light, science is not just the province of the scientist. It belongs, ultimately, to everyone. The data gathered, the methods developed, the results published: these are part of our shared inheritance. But for the public to make sense of this inheritance, to understand what is being published and under what conditions, there must be transparency, and a way to teach the medium that is scientific articles.

5.2 Manuscripts

There is a good chance that the general public's first encounter with a scientific article, one that they truly read, not merely glanced at, occurred during the COVID-19 pandemic. In that time of global uncertainty, scientific information was not merely the concern of specialists in laboratories, it spilled outward into the public domain, demanded attention, and reshaped daily lives. It was estimated that around a quarter of all new scholarly articles during that period concerned the novel coronavirus and its myriad consequences. As the world turned its collective eye toward these studies, many for the first time engaged directly with scientific manuscripts, not as students fulfilling assignments, but as citizens trying to make sense of the mechanisms of a virus that had so dramatically altered their lives.

But here we must pause, for reading a scientific paper is no small feat. It requires a kind of fluency, a set of interpretive skills that are not necessarily taught outside of specialized academic settings. One does not simply "read" a scientific article the way one might read a newspaper editorial or a historical essay. The scientific manuscript is a different kind of document: one whose surface can appear opaque or uninviting with its own set of codes, but which, when approached with the proper understanding, reveals its contents with clarity and purpose. And it is a skill worth cultivating, because the ability to navigate such documents is, in essence, the ability to participate in the great collective enterprise of science.

Let us begin by examining the anatomy of a scientific article. While the content varies by field, and the complexity of methods can differ greatly, the structure of the article is surprisingly uniform. There are, broadly speaking, three main components. The first is the scientific narrative itself, the experimental data, the analysis, and the conclusions. This is the core of the article, its substance. The second is a series of contextual and structural elements, which we might liken to metadata: the names of the contributors, their affiliations, the sources of funding, and declarations of potential conflicts of interest. The third component is a set of supplementary materials: additional data, extended discussions, or procedural minutiae that support the main text but are often housed separately, in appendices or digital repositories, and mostly made for other scientists wishing to replicate the research.

Let us look more closely at the first of these, the science itself. In previous sections, we may have examined the methods by which such research is conducted, but here, the question is how such work is presented, how one is meant to read and understand it.

The logical first step is the abstract. This short passage, usually no more than a few hundred words, is a condensed summary of the entire work. It tells you what question was asked, what method was used to explore it, and what was found. The abstract is your guidepost, your compass. It lets you know whether this article contains what you are looking for.

If the abstract appears promising, the next step, especially for those with some familiarity with the topic, is often to examine the figures. Why the figures? Because science, at its core, is a method of organizing observations of the natural world into coherent patterns. These patterns are frequently rendered visually: as graphs, diagrams, tables, or images. Each figure in a scientific paper is a distillation of many hours, sometimes years, of labor. They are designed to tell the essential story of the research at a glance. A well-constructed figure, accompanied by a clear caption, should be comprehensible even in isolation. It is no coincidence that many scientists, when drafting a new article, begin not with the text, but with the figures. These visual elements serve as narrative anchors around which the rest of the manuscript is constructed.

Consider an example. Suppose a research group has invented a new device capable of detecting the minute motion of molecules with unprecedented sensitivity. The first figure might illustrate the design of the device itself: its components, its operating principles, and its calibration. The second might compare measurements from the new device with those from existing instruments, establishing its reliability and accuracy. The final figure would likely present data from a novel application, perhaps measurements of a subtle molecular assembly never observed yet before, thus demonstrating the device's real-world utility. These three figures together tell a complete story: the inception, the validation, and the application. It is a scientific narrative in miniature.

After the abstract and figures, one might turn to the conclusion. This section distills the authors' interpretations and speculations about the implications of their findings. It answers the unspoken question: so what? What does this mean, not just for the authors, but for the field, and perhaps even the world?

For those who wish to delve deeper, or who are new to the topic, the introduction is the appropriate next step. The introduction serves as a bridge from the general to the specific. It locates the research question within a broader scientific context. Often, it will begin with a few sentences that explain the general importance of the topic. Then, it gradually narrows, citing all important previous work and highlighting unresolved questions. This process not only justifies the need for the present study but also introduces the reader to relevant literature. If you are unfamiliar with the subject matter, the introduction can be an invaluable map of the intellectual terrain.

Having explored the main body of the article, we now shift our focus to the second major component: the metadata, the scaffolding that supports and contextualizes the work. In popular culture, we often reduce great achievements to the names of singular individuals: the visionary filmmaker, the groundbreaking inventor, and the video game director. But science, in truth, is almost never the work of isolated minds. It is a collective endeavor. Scientific articles reflect this reality with a refreshing transparency.

At the top of the article, beneath the title, is a list of names who contributed to the work. And here, even the order of names is meaningful. The first author is often a graduate student or a postdoctoral researcher, the person who performed the majority of the experiments and analyses. This person has likely spent years immersed in the project. The last author is typically the principal investigator, the senior scientist who oversaw the work, obtained the funding, and provided intellectual guidance. The names in between are co-contributors: individuals who performed supporting roles in the research. In many fields, a dedicated section, often labeled "Author Contributions," spells out who did what. One person designed the experiments, another performed the simulations, and the third analyzed the results. Still, others wrote or edited the manuscript. This transparency is not only honest, it is essential. It allows us to understand the structure of scientific labor and to appreciate its collaborative nature.

Closely linked to authorship is the question of funding. Scientific research does not occur in a vacuum. It requires time, materials, equipment, and often significant financial support. At the end of most articles, you will find a section that names the institutions or agencies that provided funding. This might include government bodies, private foundations, or industrial partners. The presence of such declarations is not merely a formality. It ensures accountability. Readers have the right to know who supported the work, and whether those entities might have had a vested interest in the outcomes.

To that end, most journals also require a disclosure of any potential conflicts of interest. If, for example, one of the authors owns stock in a related company, this must be stated. These declarations help preserve the integrity of the scientific record. They remind us that scientists are human beings, operating within economic and institutional frameworks, and not immune to bias. The best safeguard against such bias is not suspicion, but transparency.

In the spirit of transparency, another crucial element of modern scientific publishing is the accessibility of data. Increasingly, researchers are required not only to describe their methods and findings but also to share the raw data underlying their conclusions. This allows other scientists to verify results, replicate experiments, or apply the data to new questions. Typically, a section will indicate where the data can be found, whether in a public repository, a supplementary file, or available upon request. This practice is more than a bureaucratic obligation. It is a cornerstone of scientific progress. Knowledge must be shareable to be useful.

Finally, most scientific articles include a short acknowledgment section. Here, the authors thank individuals who contributed to the project in ways that, while significant, did not meet the criteria for authorship. This might include colleagues who offered helpful feedback, technicians who maintained equipment, or mentors who provided encouragement. These gestures of gratitude serve as quiet reminders that science is not just a technical pursuit, but a human one. It proceeds not only by logic and data, but by conversation, collaboration, and community.

And so, a scientific article is more than the sum of its words. It is a window into a process, a method by which human beings strive to understand the universe and their place within it. Reading such an article, especially for the first time, can feel like deciphering a foreign language. But with practice, the patterns emerge, and what once seemed inscrutable becomes clear. You begin to see the structure: the framing of questions, the careful design of experiments, and the interpretation of results. You come to recognize the figures not as abstract symbols, but as expressions of empirical thought. You begin to understand not only what the authors have discovered, but how they discovered it, and why it matters.

In the end, learning to read a scientific paper is not just a technical skill but a tool to navigate life with critical thinking. How come then, from our first steps into a classroom up to our final high school years, have we never once read a scientific article in class? Why then, when learning why the sky is blue, have we never read a manuscript from Tyndall or Rayleigh?

Chapter 6
Probing Single Molecules

Abstract This chapter traces the story from early atomic theories to the cutting-edge field of single-molecule probing and its impact on drug discovery. Traditional ensemble methods, which averaged the behavior of countless molecules, have given way to techniques capable of observing biomolecular interactions one molecule at a time. These high-resolution methods reveal dynamic processes such as protein folding, ligand binding, and conformational changes with unprecedented clarity, giving critical insights for developing more targeted and effective therapeutics. The convergence of single-molecule analysis with artificial intelligence has further accelerated drug discovery, enabling screening informed by rich and massive experimental data. The second part of the chapter chronicles the historic and technical breakthroughs that led to super-resolution microscopy and single-molecule localization techniques. These innovations not only transformed imaging but reshaped our fundamental understanding of cellular processes in health and disease. As a prelude to the coming chapters on biophotonics, this narrative highlights the power of optical tools in revealing life at its most fundamental scale, and the bold thinking that drives such scientific revolutions.

6.1 Single Molecules and Drug Discovery

The study of matter at the smallest scales has always fascinated scientists, from the early atomists of ancient Greece to the great experimentalists of the twenty-first century. While philosophers like Democritus speculated about the indivisible nature of atoms, it wasn't until centuries later that technology allowed humans to begin testing those ideas. In the twentieth century, the advent of physics and advances in microscopy made it possible to not just theorize, but to observe phenomena at the molecular and atomic level. Among the transformative achievements in this scientific lineage is the field of single-molecule probing, a discipline that allows researchers to directly study individual molecules, one at a time.

This capability marks a dramatic shift in scientific inquiry. For much of history, chemistry and biology were ensemble sciences, relying on bulk and statistical measurements of billions of molecules interacting simultaneously. Even at the height

G. Graciani, *Effective Communication and Single-Molecule Research*,
SpringerBriefs in Molecular Science,
https://doi.org/10.1007/978-3-032-10426-7_6

of classical molecular biology, researchers were limited to studying large populations of molecules and drawing inferences from averaged signals. The contributions of figures like Linus Pauling in structural chemistry or Rosalind Franklin in X-ray crystallography helped illuminate molecular structures, but these insights were still derived from collective behaviors. It wasn't until the development of single-molecule techniques in the latter part of the twentieth century, facilitated by the invention of tools like the scanning tunneling microscope by Gerd Binnig and Heinrich Rohrer, and later the atomic force microscope, that researchers began probing the behavior of individual biomolecules in real time.

The precision afforded by single-molecule methods is particularly transformative in biological contexts. Proteins, nucleic acids, and other biomolecules are not rigid machines, they are dynamic, fluctuating entities that fold, misfold, bind, and dissociate in stochastic ways. These fluctuations are often lost in ensemble measurements but can be captured in minute detail through single-molecule techniques. Understanding these behaviors is critical in many areas of science, but perhaps nowhere more so than in drug discovery. A drug molecule's efficacy hinges on its precise interactions with its target, typically a protein. Subtle differences in binding affinity or conformational dynamics can mean the difference between a therapeutic agent and a toxic compound. Single-molecule studies allow scientists to observe these interactions with unprecedented clarity, capturing transient binding events, measuring forces at the pico-Newton scale, and watching in real time as a molecule alters its conformation upon binding.

This has profound implications. Single-molecule probing allows researchers to identify rare but significant events, such as a ligand that binds unusually tightly or induces an unexpected conformational change, that could be missed by conventional methods. This granular understanding can not only accelerate the discovery of new compounds but also refine them with greater specificity, reducing side effects and improving efficacy. Moreover, these methods allow scientists to investigate heterogeneous populations of molecules, acknowledging and exploiting molecular individuality rather than averaging it away.

In recent years, the landscape of drug discovery has been further revolutionized by the rise of artificial intelligence. Machine learning models, particularly those based on deep learning, now enable the prediction of molecular properties, protein structures, and even drug-target interactions with remarkable accuracy. AI can screen vast libraries of potential compounds in silico, identifying promising candidates that would have taken years to uncover using traditional methods. However, these models are only as good as the data they are trained on. Here, single-molecule probing becomes an invaluable source of high-resolution, high-fidelity data. By feeding AI systems with detailed insights from single-molecule experiments, such as real-time kinetic data, binding efficiency, and structural conformations, researchers can train models that more accurately reflect the complexities of molecular biology.

The synergy between single-molecule probing and AI-driven drug discovery represents a new paradigm. As AI becomes more sophisticated, its need for precise, mechanistic data increases. And as single-molecule methods become more automated and scalable, they can supply the kind of rich, nuanced datasets that AI thrives on. Together, these approaches offer a future where drug development is faster, more rational, and more responsive to the intricacies of human biology. In this confluence of physics, biology, and computation, the humble act of observing a single molecule has become one of the most powerful tools in modern science. What tools, then, allow for such a study? Tools that use light to probe the infinitely small.

6.2 Probing the Infinitely Small with Light

The history of microscopy stretches back over four centuries to a time when science was beginning to grasp the potential of seeing what lies beyond the limits of the naked eye. In the early seventeenth century, visionaries like Galileo Galilei and Antonie van Leeuwenhoek pioneered some of the earliest microscopes. Leeuwenhoek, with his handcrafted lenses, saw living microorganisms for the first time, describing what he called "animalcules." Over the next several centuries, advances in lens grinding, light control, and optical theory transformed microscopes from rudimentary tools into sophisticated scientific instruments. Yet by the late nineteenth century, these optical tools had reached a fundamental roadblock: what is now known as the Abbe resolution limit.

Proposed by German physicist Ernst Abbe in 1873, the limit stated that the resolution of light microscopes could not exceed about half the wavelength of visible light, around 200 nm. This meant that the intricate machinery inside cells, especially proteins and small molecules, remained blurry and inaccessible. For more than a hundred years, scientists accepted this restriction as a law of nature. While electron microscopy would later allow visualization at higher resolutions, it came with its own limitations: specimens had to be fixed, dehydrated, and often killed in the process, rendering it unsuitable for observing living biological processes in action.

In the twentieth century, advances in physics and chemistry enabled new approaches to microscopy, and with them, a string of Nobel Prizes recognized transformative progress. In 1953, Frits Zernike received the Nobel Prize for developing phase contrast microscopy, allowing better observation of living cells. Later, in 1959, Manfred Eigen shared a Nobel for studying fast chemical reactions, influencing techniques that would support later microscopy innovations. The trend continued through decades of biochemical and biophysical breakthroughs. But the great leap past Abbe's limit, the one that would truly redefine our ability to see inside life, would not come until the early 2000 s thanks to a few scientists who refused to accept the limits of conventional wisdom.

Among them was Stefan W. Hell, a German-Romanian physicist. After earning his doctorate in Heidelberg, Hell faced skepticism from his peers in Germany. Many told him that trying to overcome the Abbe limit was not just ambitious, it was futile. "You can't argue with the laws of physics," they said. But Hell was undeterred. He found refuge at the University of Turku in Finland, where a more open-minded professor gave him space to explore the impossible. There, in 1993, while reading about quantum optics, he experienced a moment of clarity, a realization that would lead to STED microscopy.

STED, or stimulated emission depletion microscopy, turned traditional assumptions on their head. Hell designed a system that used two lasers: one to excite fluorescence in molecules, and another to deactivate that fluorescence everywhere except in a nanoscopic zone. This let him scan across a sample, nanometer by nanometer, collecting data at a level of detail no one had thought achievable with visible light. His breakthrough paper in 2000 reported resolution beyond the Abbe limit, and 20 years later, Hell was appointed director at the Max Planck Institute for Biophysical Chemistry in Göttingen. The funding and freedom provided there allowed him to refine STED into a reliable and widely used imaging tool.

While Hell was breaking limits in Europe, two American scientists were independently pursuing a similar vision. Eric Betzig, once a researcher at Bell Labs, had left science disillusioned after hitting a wall in his work on high-resolution imaging. He spent some years as a stay-at-home dad before reentering the workforce in a private company. But the dream of solving the microscope's limitations continued to haunt him, echoing the concept used in Chap. 3 of passion-driven research. Eventually, he retreated to a cottage and gave himself space to think deeply, returning again and again to the Abbe problem. In 2005, inspired by new developments in fluorescing proteins, Betzig had a revelation: by switching molecules on and off and combining multiple low-density images, he could create a single, ultra-high-resolution view of a sample.

Key to Betzig's insight was the earlier work of William Moerner, who in 1997 had achieved control over individual fluorescent molecules. Working at UC San Diego at the time, Moerner's discovery of how to manipulate green fluorescent proteins laid the groundwork for Betzig's later innovation. By selectively activating a sparse subset of molecules and stitching the resulting images together, they circumvented the long-accepted optical limit. Betzig's concept earned him a position at the Howard Hughes Medical Institute's Janelia Farm Research Campus, where he built a practical tool that biologists could actually use.

In 2014, the Nobel Prize in Chemistry was awarded jointly to Hell, Betzig, and Moerner. Their work didn't just refine existing methods, it opened an entirely new field: nanoscopy. For the first time, scientists could observe proteins, organelles, and even single molecules inside living cells with unprecedented clarity. Today, STED and single-molecule microscopy are used around the world to study everything from brain synapses to cancerous cells, from the early stages of embryonic development to the mechanisms of neuro-degenerative diseases like Alzheimer's.

Stefan Hell's path to that Nobel moment was marked by resistance, isolation, and doubt. Yet what fueled him was not mere ambition, it was a refusal to accept that the visible universe stopped where traditional optics failed. His story, like those of Betzig and Moerner, is a reminder that science often advances not by reinforcing what is known, but by challenging it. From the handmade lenses of Leeuwenhoek to lasers scanning living cells, the microscope has become more than an instrument, it's a symbol of what human vision, fueled by persistence and creativity, can achieve.

Being myself a researcher focused on developing new optical-based sensing techniques and instruments, the next chapters will be focusing on how photonics and biology interact in the field of biophotonics.

Chapter 7
Single-Molecule Biophotonics

Abstract This chapter explores the scientific journey behind understanding why the sky is blue, beginning with John Tyndall's pioneering experiments on light scattering and culminating in Lord Rayleigh's theoretical framework grounded in Maxwell's equations. It introduces the Tyndall effect and explains how Rayleigh scattering by air molecules accounts for the blue hue of the sky. The discussion then transitions into modern applications of light scattering in biology, particularly the challenges posed by the small size and low scattering efficiency of biomolecules. To overcome these limitations, the chapter presents three biophotonics strategies: (1) using scattering markers like DNA-tethered beads to produce detectable signals, (2) employing advanced image processing techniques to detect and study proteins without labels, and (3) amplifying weak scattering through cavity-based interferometric techniques. These methods enable sensitive analysis of protein behavior and interactions, crucial for advancing fields like drug discovery and precision medicine. The chapter illustrates how foundational physics informs state-of-the-art biological measurements, and how innovations in light manipulation continue to push the boundaries of what we can observe in the molecular world.

7.1 Blue Sky

7.1.1 The Question

Why is the sky blue? This universal question, spoken from the mouths of every child, has puzzled natural philosophers for centuries. The very ceiling of our world, painted daily in hues of blue, contains within it clues about the structure of matter, the behavior of light, and the very composition of the atmosphere we breathe.

In the late1860 s, it was this simple question that arrested the curiosity of Irish physicist John Tyndall. He was not alone in his fascination, but unlike the casual sky-gazer, Tyndall pursued the answer with the tools of science and the unshakable belief that through experiment and reason, Nature might be coaxed into revealing its secrets.

G. Graciani, *Effective Communication and Single-Molecule Research*,
SpringerBriefs in Molecular Science,
https://doi.org/10.1007/978-3-032-10426-7_7

Yet, as with all true scientific discovery, the roots of his question were entangled in deeper, practical concerns. Tyndall was interested in how particles suspended in the air influenced light propagation. His inquiries into air purity and germ transmission, far from being distractions, were the very soil from which his work on light scattering grew. It was a time when understanding the unseen could mean the difference between health and plague, between foggy ignorance and clear reason.

7.1.2 The Experiment

To simulate the vast and ever-changing atmosphere in the controlled confines of a laboratory, Tyndall designed a most elegant and instructive device: a long, transparent glass tube, sealed at both ends, through which he could pass light and introduce controlled impurities.

Into this tube he sent a beam of white light, a stand-in for sunlight, and introduced microscopic particles. Smoke, fine dust, or water droplets condensed from vapor. These were not arbitrary choices. They mimicked the tiny particles suspended in the atmosphere, whose role in modulating light he sought to understand.

As he observed the beam through the sides of the tube, something astonishing happened: the path of the light began to glow with a striking blue hue. The color resembled that of the daytime sky. And then, looking toward the beam's origin from the far end of the tube, he saw something else: a warm reddish glow, reminiscent of the setting sun.

Tyndall had created, in this modest apparatus, a laboratory version of the sky itself.

More telling still was what happened when he cleaned the tube of all particles. The light, deprived of anything to scatter off, disappeared from view when looked at from the side. The message was unmistakable: it was the particles suspended in the air that scattered the light and made visible its blue color.

7.1.3 The Understanding

From these experiments arose what we now call the Tyndall effect: the scattering of light by particles small enough to remain suspended in a medium, yet large enough to interact with light in complex, colorful ways.

Tyndall was not merely a conjurer of beautiful effects. He was rigorous. He altered the size and type of particles, demonstrating that finer particles produced deeper blue scattering. He kept meticulous records and ensured that his demonstrations were repeatable. He isolated variables, adjusted conditions, and let the data speak.

The blue sky, he concluded, was the result of sunlight scattering off small airborne particles, perhaps dust, perhaps condensed droplets of water. The precise identity of these scatterers remained elusive, but the principle had been planted. The way was now open for others to explore.

7.1.4 The Collective Enterprise

In the decades that followed, another scientific titan took up the mantle. Lord Rayleigh was not content with a qualitative explanation. His ambition was greater: to place the blue of the sky within the rigorous framework of electromagnetic theory, to derive its properties not just from experiment but from the mathematics of Maxwell's equations themselves.

While Tyndall worked with visible particles, smoke and droplets, Rayleigh turned his attention to the minuscule. He focused on particles far smaller than the wavelength of light itself: molecules. Here, in the realm of the very small, light behaved differently.

He extended his analysis to molecular suspensions and, more audaciously, to the air molecules themselves. What if the very fabric of the atmosphere, its constituent gases, were sufficient to explain the scattering? The answer was yes.

Rayleigh found that the intensity of scattered light increased sharply as the wavelength decreased. In practical terms: blue light (short wavelength) is scattered far more than red light (long wavelength). His measurements showed that the scattered intensity varied inversely as the fourth power of the wavelength.

This meant that blue light, wavelength about 450 nm, is scattered nearly ten times more than red light at 700 nm. And violet? Even more. But our eyes are less sensitive to violet, and some of it is absorbed by the atmosphere. What we see is the sum of physical laws and a biological response to it.

In 1871, Rayleigh published his masterwork: "On the Light from the Sky, Its Polarization and Colour" [1]. Therein, he married experiment with theory, geometry with optics, observations with mathematics. It was the culmination of years of careful thought and meticulous work.

Rayleigh's methodology is a textbook case of scientific practice at its best. He formulated hypotheses grounded in Maxwell's theory of electromagnetism. He conducted controlled experiments with molecules and laboratory suspensions. He looked up to the sky and checked his predictions.

And when his theory didn't quite fit the data, he refined it. Over time, he incorporated both wavelength dependence and polarization. By 1899, he had extended his model to explain the hue and brightness of the sky with stunning accuracy.

He did something more profound still: he showed that the molecules of air, oxygen and nitrogen, were enough to scatter light. No need for dust or droplets. The sky's blue was an intrinsic feature of the atmosphere's molecular structure. It is the very same effect that biophotonics labs use today to study single molecules.

7.2 Light Scattering

At the heart of this story is a deeper understanding of what light scattering actually is.

When a beam of light encounters a particle, whether a molecule of air or a speck of dust, it perturbs the electrons within it. These charged particles, agitated by the oscillating electric field of the light wave, themselves begin to oscillate in unison. And as they do, for all that they are charged particles, they generate and radiate light of their own in new directions.

This emission of light is what we call light scattering.

There are actually many kinds of scattering, which depend on the fine details of this light-matter interaction, and two actually stand out. Elastic scattering, such as Rayleigh's, preserves the wavelength of light: the color that comes in is the same that goes out. Inelastic scattering, like Raman scattering, operates a change in energy, while the excited oscillation occurs: the molecules, due to their internal structure, can absorb some of it. The result: a less energetic light emission, and a change in the re-emitted wavelength itself.

But not all particles scatter light equally. To quantify how strongly a particle interacts with light, we use a concept called the *scattering cross-section*. It is not just a physical area as the name implies, but an effective one: a measure of how much of the light a particle can intercept and interact with.

The cross section depends on many factors: the actual size of the particle compared to the wavelength of course, but also the refractive index contrast between the particle and the medium in which it is. If a particle has a refractive index close to that of its medium—think a bead of glass in water—light won't be able to *see* it, to interact with it. And there is more: those properties depend on the wavelength of light.

A grain of dust has a large cross section and scatters copiously. A molecule of oxygen, far smaller, has a tiny cross section, yet the air is full of such molecules, and their cumulative effect paints the sky.

Light is an exquisite probe. It travels great distances, can reveal the structure of atoms. It must therefore be a formidable tool to probe biological molecules? Yet, for all its versatility, it has limits. When it comes to proteins, nucleic acids, lipids, light's powers fade.

Why? First, the scale. These molecules are measured in nanometers, ten to a hundred times smaller than the wavelength of visible light. In this regime, scattering becomes feeble. Second, contrast. Proteins are made of common elements, all of which are similar in optical properties to water, the medium in which they typically evolve. The refractive index of water is about 1.33; for proteins, it's close to 1.59 on average. The difference is small. The result: light barely "notices" the boundary between two.

The implications are profound. The scattered light from individual proteins is not just weak, it is drowned out by background noise. A single speck of dust in a sample scatters more light than millions of proteins. A stray bubble or impurity can render a delicate measurement meaningless.

Even when we can detect scattering, interpreting it is a mathematical ordeal. The signal is indirect. We do not see the molecule, we infer its shape and properties through complex inversions.

Yet we do not give up. Innovations in nanophotonics, clever tricks with interference, image processing algorithms or amplifying cavities are all new strategies to peer into the molecular world through light scattering.

7.3 Biophotonics

7.3.1 Strategy 1: Scattering Markers

We have previously seen that most biological objects relevant to our interest do not interact much with light. The first solution to that issue is simply to use chemistry in order to attach a strong scattering marker to objects of interest.

In essence, the experimental setup is simple. We attach a scattering microscopic bead to the protein under study, and the protein to the surface of a microscope coverslip. This is done, of course, in a more controlled manner: molecular tags and spacers are used. To provide spatial separation between the bulky magnetic bead and the protein, and to reduce nonspecific interactions, DNA spacers are often employed. These are double-stranded DNA molecules of known length and mechanical properties that serve as rigid linkers. One end of the DNA spacer is attached to the protein, while the other end is bound to the bead. One commonly employed attachment method involves the use of the SpyTag/SpyCatcher system. SpyTag is a short peptide that forms a spontaneous covalent bond with the SpyCatcher protein domain. By genetically engineering the protein of interest to include a SpyTag at both its N- and C-terminals, and covalently linking a SpyCatcher protein on the DNA spacer, a stable and specific attachment bridge can be established.

The beads are a few microns large (much bigger than the protein) and are therefore easy to see with a traditional microscope using light scattering. By studying the motion of those trackers, we can infer in minute details the motion of the protein. However, since we are interested in studying the native dynamic behavior of proteins, one can rightfully argue that those big cumbersome objects hinder their motion. Figure 7.1 shows what such an attachment might look like.

Those attachments are used for example in magnetic tweezers experimental setups, to study the conformational dynamics of proteins, how they fold and sometimes misfold. This will be detailed as an example in the next chapter.

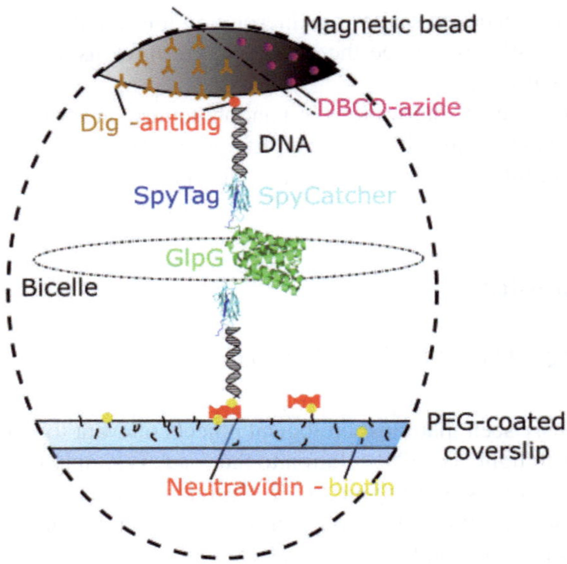

Fig. 7.1 Schematic diagram of a magnetic bead scattering marker attachment scheme for studying the GlpG membrane protein. The protein is inserted into a lipid bicelle and anchored between a PEG-coated coverslip and a magnetic bead through DNA spacers and specific chemical linkages. The DNA is attached to the protein via a SpyTag/SpyCatcher system. The top of the DNA is connected to the magnetic bead through either a digoxigenin–antidigoxigenin interaction or DBCO-azide click chemistry, while the bottom is anchored to the surface of the coverslip using a biotin-neutravidin bond

7.3.2 *Strategy 2: Image Processing*

Mass photometry is a cutting-edge technique used to study proteins and their interactions at the single-molecule level, without the need for scattering labeling [2, 3]. As explained before, one of the fundamental challenges in visualizing proteins with light is that they do not scatter strongly enough to be easily seen under normal conditions. However, mass photometry overcomes this limitation through highly sensitive image processing and interference-based detection schemes.

The core idea behind mass photometry is to measure the light scattered by a protein the exact moment when it lands on a glass coverslip. Even though the signal from a single protein is extremely weak, the system uses a special type of microscope called interferometric scattering microscopy (iSCAT), which can detect minute changes in a light signal. When a protein binds to the coverslip, it creates a tiny but measurable change in the intensity of the image in the form of an interference pattern. Sophisticated algorithms are then necessary to detect and track individual molecules as they land and bind on the coverslip. Figure 7.2 reveals the principle behind this detection: a movie, constituted of individual frames, is constantly being recorded. Most of the

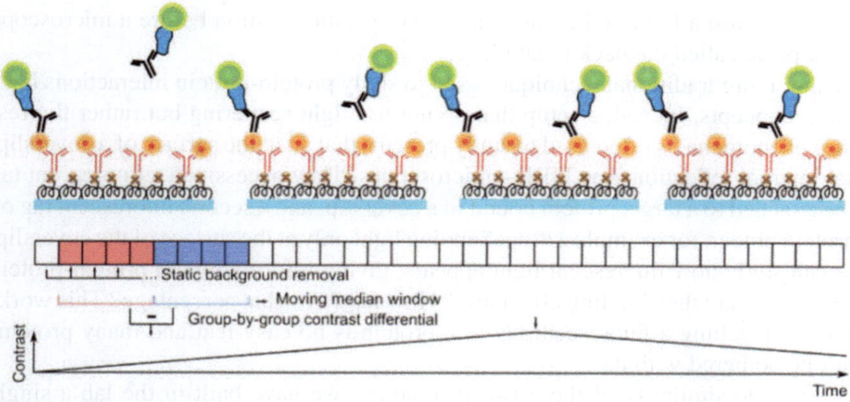

Fig. 7.2 PEG-coated and passivated microscope coverslip with red-fluorescent tagged antibody bound to it. A green-fluorescent tagged protein approaches the surface and binds to the target antibody. Frames of a movie are recorded. A moving median window as well as static background removal are performed before subtracting two adjacent groups of frames. Resulting local contrast is measured and plotted over time

time, nothing happens, and the frame is blank. Well, actually not blank, but filled with all sorts of noise coming from the setup: stray light, laser reflections, and so on. The algorithm is essentially always taking two adjacent frames and subtracting one from the other. If a protein created a tiny intensity change in the image, it will be revealed as the noise disappears when the subtraction happens. Everything static in the image fades, leaving only the new local information: the protein. In practice, we do not do it with only two frames, be we are constantly averaging a few frames in order to get rid of other types of dynamic noise. Now imagine a protein is slowly approaching the coverslip surface, the signal—it's actually a contrast that we measure—will slowly increase until it reaches a maximum when the object sits on the surface and is in full view. Once the protein is here, it is not moving anymore, hence the frames are virtually identical for the rest of the film. The algorithm keeping on doing the same subtraction will result in a contrast slowly fading over time as shown in Fig. 7.2.

What makes this technique even more powerful is that the amount of scattered light directly relates to the mass of the protein. By calibrating the system using proteins of known mass, it becomes possible to determine the molecular weight of unknown particles. This means that not only can we see individual proteins but also can measure their mass with high precision. When studying protein-protein interactions, mass photometry allows researchers to detect binding events by observing shifts in mass as two or more proteins come together to form complexes. By measuring how often such binding events occur and under what conditions, researchers can infer binding affinities, stoichiometries, and interaction dynamics.

This effect has two very important properties. First, the detection takes place right at the surface of the microscope coverslip. Second, in order to properly lit the sample, one need to use a laser and focus it at a very specific location before a microscope lens, a place called the back focal plane.

Other more traditional techniques used to study protein-protein interactions have similar concepts. Indeed, a setup that do not use light scattering but rather fluorescence phenomena, is often used to study proteins that sit at the surface of a coverslip: total internal reflection—or TIRF—microscopes. They necessitate a fluorescent tag to be attached to a target protein bound to a coverslip, and a second fluorescent tag on a protein binder, for example a drug. Sending light only at the surface of the coverslip, one can study how fluorescent light appears, giving information on protein-protein interactions: are they binding efficiently? How fast? At what percentage? This works well but attaching a fluorescent tag to a protein is no easy feat and many proteins can't be bothered with it.

Given the similarity of these two techniques, we have built in the lab a single setup that combines both. Figure 7.3 shows how we achieve that. We send three lasers, one for scattering, or mass photometry, and two for fluorescence microscopy. Given the similarity of the setups, the three lasers are combined and follow the same path, allowing for a compact setup. Only the blue laser, used for mass photometry, is focused at the back focal plane of a lens, while the other are focused a little bit off axis. This allows us to study protein-protein interactions under every possible aspects: carefully monitoring not only how well they bind using TIRF fluorescence but also how well the tagging is made using the scattering signal from mass photometry. This technique is an excellent example of how state of the art experimental instruments are made in the lab today: by small increments and minute ameliorations to never stop reaching farther.

7.3.3 *Strategy 3: Signal Amplification*

7.3.3.1 Why?

Researching how to develop more effective and targeted treatments for diseases, particularly cancer, one area of biology has become increasingly important: protein-protein interactions (PPIs). These interactions are the foundation of nearly every cellular process, from signaling pathways to immune responses. When two proteins bind together, they often trigger or regulate essential biological functions. Disruptions in these interactions are at the root of many diseases, including cancer, neurodegeneration, and autoimmune disorders. By understanding and manipulating these interactions, scientists can design highly specific drugs that precisely target malfunctioning pathways without harming healthy cells. One of the most promising applications of this principle is in the development of antibody-drug conjugates (ADCs), a cutting-edge class of therapeutics that combines the specificity of antibodies with the potency of small-molecule drugs. These are engineered molecules that link a monoclonal antibody to a drug. The antibody portion of the ADC guides the drug to a specific target, often a protein on the surface of cancer cells, while the toxic

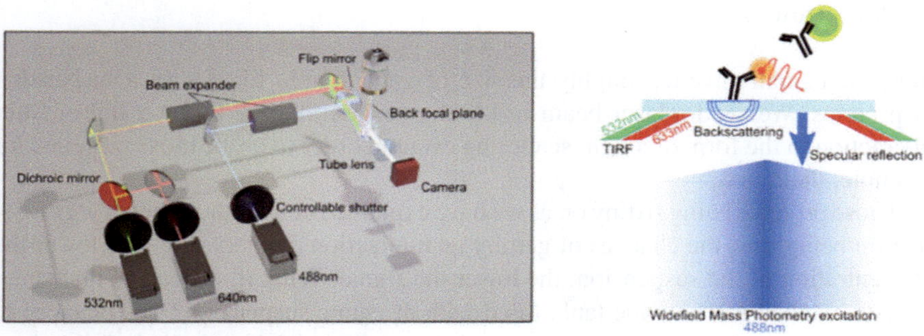

Fig. 7.3 (left) Optical setup for multi-color laser illumination in advanced fluorescence or scattering microscopy. Laser beams of 488 nm (blue), 532 nm (green), and 640 nm (red) are individually controlled by shutters and combined using dichroic mirrors. The resulting multi-wavelength beam is expanded and directed into the back focal plane of a high numerical aperture objective using a flip mirror. The light interacting with the sample is collected and focused through a tube lens onto a camera for image acquisition. This configuration allows for rapid switching between a mass photometry experiment with the 488 nm laser being focused at the center of the back focal plane while the other lasers are focused off-axis, allowing for a TIRF experiment. (right) Microscope coverslip illumination scheme allowing for both a widefield 488nm illumination producing a backscattering signal of objects attached to the surface and 532 nm and 633 nm TIRF excitation for fluorescent dyes close to the surface of the coverslip

payload is designed to kill those cells once the ADC is internalized. The success of an ADC hinges on a deep understanding of PPIs: the antibody must bind tightly and selectively to its target protein, avoid off-target interactions, and remain stable until it reaches its destination.

Studying PPIs is therefore critical to improving ADC design. For example, researchers need to understand which proteins are uniquely or over-expressed on cancer cells but not on healthy cells. This requires mapping the interaction networks and expression profiles of different cell types. Once a suitable target protein is identified, the next challenge is designing an antibody that binds specifically to it, and only to it. This involves fine-tuning the antibody's binding regions so that they interact precisely with the target protein's surface. Today, this means testing the binding efficiency to a target of many different versions of an antibody. Practically, slight variations in sequences are made, molecules are produced and tested against targets, data accumulated and fed to a custom neural network made to calculate the best possible molecule.

This approach holds immense promise for treating diseases that have long been considered difficult or impossible to tackle with traditional pharmaceuticals. As we move further into the age of precision medicine, mastering protein-protein interactions will be key to unlocking a new generation of life-saving therapies. In vitro tests are unfortunately far from the complex processes that happen in the body. It is therefore essential to have tools that are able to quantify those interactions in environment native to those proteins, with minimal external perturbations or markers. This is where light scattering can play its card right.

7.3.3.2 How

Imagine a small cuvette, roughly the size of your thumb, filled with a suspension of proteins. We send a laser beam inside the cuvette, hoping to get a light-matter interaction in the form of a light scattering event in order to probe those proteins, for example, their size.

Those proteins being so tiny or, as we have explained before, their scattering cross-section being tiny, the chances of getting an interaction are very low. The lower the concentration of the suspension, the lower the signal. What if, instead of that small cuvette, we had a 10 m wide tank, filled with the same suspension? The chances of producing a meaningful interaction would be much higher. It is unrealistic for obvious practical reasons. What if instead the suspension was much more concentrated? Also unrealistic because it is hard and costly to produce large quantities of such samples.

The solution: a general method called 3D stochastic interferometry [4] and, more importantly, one of its applications to probing a suspension: Cavity Amplified Scattering Spectroscopy [5]. The idea is simple: by placing the sample inside a highly reflective cavity, the chances of getting a light-matter interaction are multiplied. If a photon does not interact with a protein as it goes through the sample, it gets another chance to probe the medium after it is reflected back into it. Typically, thousands of reflections occurs inside the cavity. Everything is happening as if the sample was probed thousands of times or, equivalently, as if its size was a thousand times bigger. A 1 cm cuvette is virtually replaced by a 10 m long tank without the need to ever change the sample itself.

It has been shown that such a technique can unlock the measurement of protein dynamic through their scattering signal alone, without the need of external markers [6], which makes it extraordinarily good at studying the native behavior of those objects. In the context of AI used for drug discovery, this method is very promising because it is able to deliver great amounts of data in short amounts of time through mass throughput experiments, requiring minimal sample preparation, a necessary condition to train machine learning models.

This technique, first invented in the lab as a way to study proteins, is now in the process of being industrially developed in other adjacent contexts: for drug discovery, vaccines stability monitoring, but also water quality testing and the quantification of nanoplastics in the oceans, showing that the reach of techniques developed in the lab is often wide and unexpected.

References

1. Rayleigh, L.: On the light from the sky, its polarization and colour. Phil Mag. **41**, 274 (1871)
2. Young, G., Hundt, N., Cole, D., Fineberg, A., Andrecka, J., Tyler, A., Olerinyova, A., Ansari, A., Marklund, E.G., Collier, M.P., et al.: Quantitative mass imaging of single biological macromolecules. Science **360** (2018)
3. Graciani, G., Yoon, T.Y. : The new kid on the block: mass photometry. Mol Cells. **46**(3), 187–189 (2023)

4. Graciani, G., Filoche, M., Amblard, F.: 3D stochastic interferometer detects picometer deformations and minute dielectric fluctuations of its optical volume. Commun. Phys. **5**(1), 1–10 (2022)
5. Graciani, G., King, J. T., Amblard, F.: Cavity-amplified scattering spectroscopy reveals the dynamics of proteins and nanoparticles in quasi-transparent and miniature samples. ACS Nano **16**(10), 16796–16805 (2022)
6. Graciani, G., King, J. T., Amblard, F.: Marker-free protein study by amplified light scattering. In: SPIE Advanced Biophotonics Conference (SPIE ABC 2021), vol. 12159, pp. 31–36. SPIE (2022)

Chapter 8
Real-World Application to Single-Molecule Research: Protein (mis)folding

Abstract This chapter sheds light on the vital world of protein biogenesis, folding, and misfolding, processes at the heart of cellular life. Proteins, the molecular machines of life, must fold precisely into three-dimensional structures in order to function. Missteps in this delicate choreography can lead to severe diseases such as Alzheimer's, Parkinson's, or cystic fibrosis. Cells deploy sophisticated quality-control systems, chaperones, and degradation pathways to manage this complexity. Yet, environmental stress, mutations, or aging, can overwhelm these safeguards. Membrane proteins, in particular, pose unique challenges: they comprise nearly a third of human proteins and serve as critical receptors and transporters, yet their complexity renders lab study difficult. Investigating how these proteins fold, insert into membranes, and sometimes misfold, is essential for understanding disease mechanisms and advancing drug development and synthetic biology. In our laboratory, we study these systems using custom-built magnetic tweezers to measure forces at the pico-Newton scale. This requires a completely multidisciplinary expertise, covering large facets of science: optics, engineering, biochemistry, and computer science. We describe here the experimental design, noise-suppression strategies, and high-resolution, high-frequency measurements that we can perform on proteins. Through meticulous instrumentation and data analysis, we reveal detailed folding pathways and dynamic intermediates, demonstrating how state-of-the-art single-molecule techniques unlock new insights into protein behavior, with profound implications for new therapies and biotechnology.

8.1 Why?

Proteins are the molecular machines that carry out virtually every function in living cells. From catalyzing reactions to transmitting signals and maintaining structural integrity, their roles are vast and essential. However, to perform these tasks effectively, proteins must first be accurately synthesized and correctly folded into their unique three-dimensional structures. When this process, known as protein biogenesis and folding, goes awry, it can have profound consequences for cellular health and lead to a wide range of diseases. Among the diverse categories of proteins, membrane proteins

G. Graciani, *Effective Communication and Single-Molecule Research*,
SpringerBriefs in Molecular Science,
https://doi.org/10.1007/978-3-032-10426-7_8

stand out for their complexity, functional significance, and the unique challenges they present. Understanding how these proteins are made, folded, and inserted into cell membranes, is a vital area of research that bridges basic biology, medicine, and biotechnology.

Protein biogenesis begins with the translation of genetic information encoded in DNA into chains of amino acids. These chains must then fold into precise shapes to become functional proteins. This folding process is not random, it is guided by both the protein's sequence and a sophisticated cellular infrastructure that includes chaperones and quality control mechanisms. Failures in folding can result in proteins that are nonfunctional or even toxic. Misfolded proteins may aggregate, forming clumps that disrupt cellular functions. Such aggregates are hallmarks of neurodegenerative diseases like Alzheimer's, Parkinson's, and Huntington's. By studying how proteins fold and what causes them to misfold, scientists can gain insights into the root causes of these conditions and potentially identify new therapeutic targets.

Despite the importance of proper folding, the cellular environment is inherently crowded and dynamic, making it a remarkably delicate and error-prone process. Cells mitigate this risk to ensure faulty proteins are either refolded or degraded. Yet, the system is not foolproof. Aging, genetic mutations, and environmental stress can tip the balance, allowing misfolded proteins to accumulate. Understanding the intricacies of this process is critical not only for developing treatments for protein misfolding diseases but also for designing better biotechnological tools, such as more efficient production of therapeutic proteins.

Membrane proteins, which are embedded in the lipid bilayer membranes of cells and organelles, add another layer of complexity. They represent roughly one-third of all proteins encoded in the human genome and perform vital roles, including transporting molecules, transmitting signals, and catalyzing chemical reactions. Many of the body's most important receptors, including heat receptors or ion channels, are membrane proteins. Because of their central role in cellular communication and transport, they are also the target of more than half of all current pharmaceuticals.

Despite their importance, membrane proteins are notoriously difficult to study. Their folding is not merely a matter of achieving the right structure but also of proper insertion and orientation within the cell membrane. In laboratories, it is extremely difficult to study them directly inside this native environment.

The next section describes in details how this can be done experimentally in the lab today.

8.2 How?

8.2.1 Inside the Lab

For this kind of state-of-the-art setup, there is no equipment you can buy. You have to do it all by yourself: buying mechanical parts and optical lenses, actually building the setup and isolation, doing the chemistry to produce the sample and, finally, performing the experiment. This is not an easy task that goes with constant refinement, control, and repairs. This kind of work therefore requires a multidisciplinary team: biochemists to find the right way to attach markers to proteins, expert in experimental optics to build the setup, engineers to code for the whole experiment automation and data processing.

8.2.2 *Introduction*

The ability to measure forces and displacements generated by single molecules brought formidable insights to fundamental biological processes ranging from the study of cellular motility [1, 2] to protein folding pathways, or even the replication of DNA [3]. Those molecular forces are typically operating at the nanometer and pico-Newton scales, and drive virtually all biological motion. In only a few decades, measurement techniques tailored and optimized for those detection windows went from their first development to now being routinely used in bio-engineering labs. In particular, three techniques are now recognized as the building blocks of single-molecule spectroscopy: optical tweezer (OT), atomic force microscopes (AFM), and magnetic tweezers (MT). This section describes a super-resolved magnetic tweezer framework, including both hardware and software setups, together with noise correction schemes, data processing algorithms, and bio-chemical pulling constructs engineering, all designed to study the insertion and folding of membrane proteins.

There is a long series of protocols fine-tuning optical tweezers and atomic force microscopy setups for the faster and better-resolved study of various single-molecular systems. Ranging from direct living cell studies [4, 5] to biochemical reactions monitoring [6] or work on membrane proteins [7], AFM can now produce high-speed measurements [8, 9] and have direct imaging capabilities [10]. In addition, relying on complex combinations of a variety of techniques, such as microscopy or Raman spectroscopy, optical tweezers protocols have now become highly technical and sophisticated [11]. Even though they proved to be very versatile, OT and AFM suffer a few limitations, some of which could potentially be overcome by the method described in the present section.

First, being a light-based technique, optical tweezers necessarily interact with any dielectric material under probe, and thus lack selectivity: the photon radiation pressure not only affects the optically sensitive probe particles, but any molecule present in the laser focus. Furthermore, since the forces originate from light intensity gradients, they can easily be perturbed by the sample itself if it is not highly homogeneous. In practice, any change in the local refractive index density of the sample will affect the light path and may reduce the trapping efficiency. Finally, the high intensity at the laser focus required for trapping results in local heating that may be significant and affect the sample [12].

Second, atomic force microscopes often better perform in high-force applications. They are limited in their interactions by the size and stiffness of their cantilevers, which are often introducing spurious interactions that are difficult to disentangle from the specific binding of the studied molecules.

Magnetic tweezers, on the contrary, are the simplest setup of the three. They can exert forces that are highly selective to the probe magnetic beads, introducing virtually no perturbation to the sample. The magnetic gradients produced are nearly constant at the scale of a few microns, can be used to produce large displacements and rotations, and can exert forces as low as a few pico-Newtons. Those specificities allow for a wide range of applications for delicate and possibly heterogeneous samples, from the study of biopolymers mechanics [13], the dynamics of motor proteins [14], or the (dis)-association of biomolecular complexes involved in mechanobiological processes [15] to, as detailed here, the study of protein folding pathways. Nonetheless, magnetic tweezers have their own share of limitations. They lack the manipulation ability of the aforementioned techniques but, more importantly, are difficult to use in high-frequency settings. Indeed, they rely on video feeds for 3D tracking and are thus often limited to frequencies in the kilo-Hertz range (2.4 kHz in the present protocol). Better performing at low frequencies and long acquisition times, they are especially sensitive to noises, dominantly drift phenomena. In particular, the present protocol introduces specific schemes to reduce those noises and achieve super-resolution (0.3 nm at 1.2 kHz, averaged over 5 Hz window).

Initially used for mapping the energy landscape of the second-stage folding of single membrane proteins at a mid-frequency range [16], the setup presented in this section has quickly been improved to work at higher frequencies, up to 1.2 kHz. Making use of the specific clamping ability of magnetic tweezers, it was then implemented for the study of single neuronal SNARE complexes formation under high mechanical tension [17] and their disassembly [18]. Taking full advantage of the finely tuned forces allowed by magnetic tweezers and the possibility to work at high frequencies, direct observations of protein folding revealed specific folding pathways of membrane proteins [19] and, recently, a balance between foldability and functionality of glucose transporters [20].

As an example, we will explore here the schemes used for the study of the folding pathway of the Escherichia coli rhomboid protease GlpG.

8.2.3 Experimental Setup

The experimental setup is straightforward and detailed in Fig. 8.1. A collimated super-luminescent diode illuminates a sample placed on the stage of an inverted microscope. The light first passes through a 1 mm space separating a pair of permanent magnets controlled for both vertical translation and rotation around the optical axis. A 100x oil-immersion objective complemented by a high-speed CMOS camera form an image grabbing scheme working at 4 kHz. The objective itself can see its vertical position controlled by a piezo-actuator used for both auto-focus and calibration in conjunction with a tracking algorithm. Because all the tracking schemes rely on analyzing interference patterns from magnetic beads, the light source needs to be

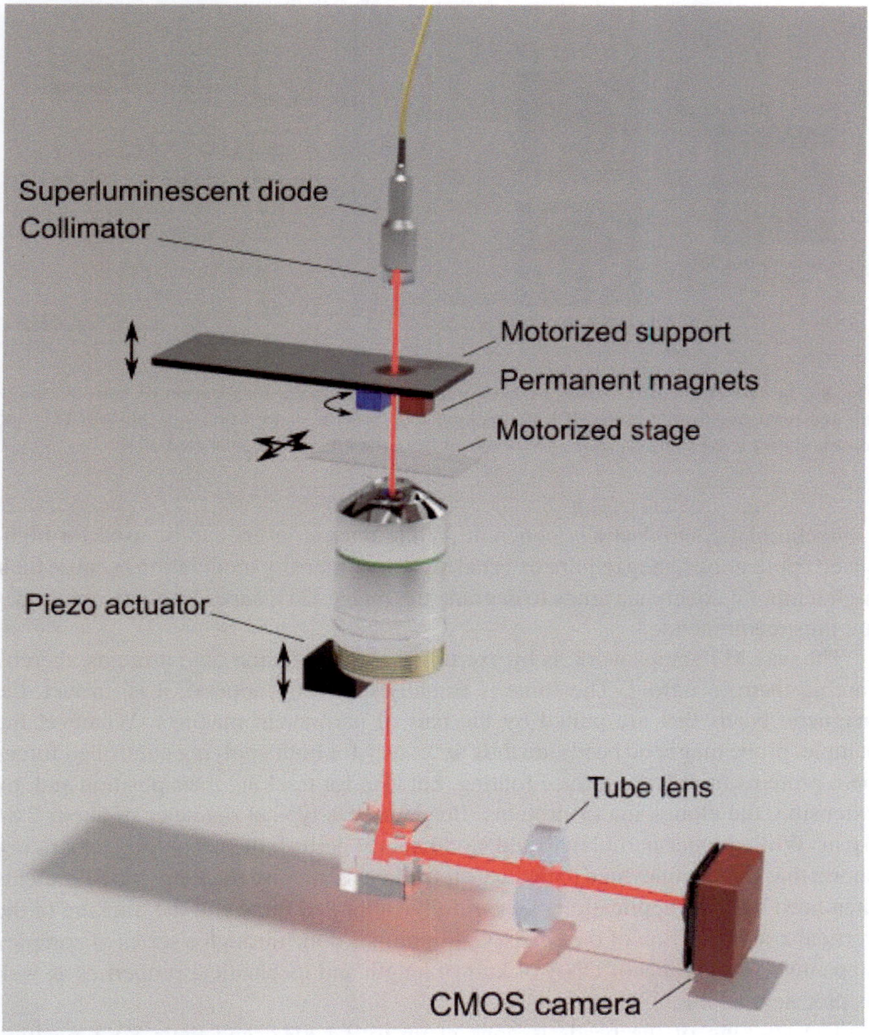

Fig. 8.1 Schematics of a magnetic tweezer experimental setup. A collimated super-luminescent diode illuminates a sample placed on a XYZ motorized stage. A Z-motorized objective lens produces an image through a tube lens onto a CMOS camera. A pair of Z-motorized permanent magnets are placed above the sample

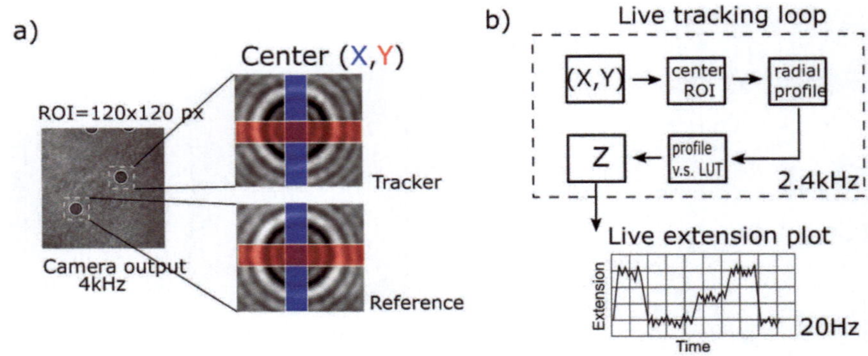

Fig. 8.2 **a** 120 × 120 pixels camera output image and automatic localization of both the tracking and reference bead's center. **b** Live tracking loop process using Look-Up-Tables (LUT) and representative live extension time trace obtained from tracking and represented 20 Hz

monochromatic and spatially coherent. While intense lasers can be used for high-speed applications, they require external temporal intensity modulation because their high temporal coherence tends to degrade the image [21]; super-luminescent diodes are thus recommended.

The way MT setups work, is by exerting a pulling tension onto proteins, thereby forcing them to unfold. The force is actually indirectly applied, it is, in fact, the magnetic beads that are pulled by the pair of permanent magnets. Whatever the sample, probe magnetic beads are thus necessary for both applying controlled forces onto proteins in order to induce folding, but also for tracking their position and, by extension, the elongation of proteins. Beads have a typical diameter of about 2.8–3 μm. While bigger particles would significantly perturb the unfolding, experience shows that smaller ones tend to fluctuate too much for precise tracking. Measurements then need both the application of a carefully calibrated force and the tracking of the vertical z-axis position of the beads. Calibration is done through a series of complex steps involving strains of DNA of known length and mechanical properties as well as precise tracking.

The tracking of the beads is done at up to 2.4 kHz, and their XYZ position graphs are plotted and updated in real time 20 Hz. Most of the fine analysis of the displacement traces are done after the fact, it is however important to note that having access to real-time traces is crucial to identify pulling construct defects (double and/or off-center attachment, beads stuck on the slide, etc.). Also, in the case of complex measurements, such as protein folding, it is essential to observe how the protein folds and unfolds in real time in response to the pulling force in order to adapt the force profile accordingly.

The tracking (Fig. 8.2) of the extension of proteins is actually done through the analysis of the diffraction pattern of the magnetic beads that are attached to them. More precisely, we use the ring-shaped interference pattern resulting from the interference of both the incident light wave propagating through the sample up to the microscope objective, and the wave forwardly scattered off the beads. Said pattern being highly dependent on the distance of the bead from the objective (its position on the vertical z-axis), its position and, by extension, the state of the protein, can be recovered through an image analysis scheme. In practice, the tracking is done in two steps: 1. The x-y position of the center of the bead on the plane normal to the optical axis is obtained (Fig. 8.2 a)) and 2. after centering a region of interest (ROI) around the center position of the bead, the interference pattern is analyzed to recover its z-position (Fig. 8.2 b)).

1. Two profiles calculated from vertical and horizontal stripes averaged over a width of 10 pixels are used to find the bead's center. The profiles are auto-correlated and the bead center is found at the maximum correlation position, as interpolated by quadratic fitting.
2. In practice, two classes of tracking algorithms are routinely used: Look-Up-Tables (LUT), which rely on a direct comparison of the bead's pattern to a library of calibrated images at known positions and holographic methods that recover the position of the bead by direct calculation and simulation of the interference pattern using light scattering theory. The most commonly used theoretical derivations rely on the Lorenz-Mie Scattering Theory (LMST) and the Rayleigh-Sommerfeld back-propagation [22] although other less-used techniques such as Fresnel-Kirchhoff or Fourier propagation exist [23].

For offline analyses, the force profile (obtained from the magnet position and the calibration curves), the XYZ beads position time traces, the sampling time, the beads parameters (radius, off-center attachment, etc.), and, eventually, the position or the objective are exported. We should point out that measurements typically require two beads: a tracker attached to the protein of interest and a reference bead immobile on the microscope slide. Both their Z positions are exported, and the tracker signal is used after subtracting the reference trace in order to help reduce instrumental drift.

8.2.4 Noise Suppression

Performing super-resolved measurements requires a fine knowledge of the sources of noise affecting the system and systematic ways to overcome them. The main extrinsic sources of noise are the acoustic and seismic vibrations (including from internal parts of the system, such as motors or piezo-actuators), thermal fluctuations, stray light, and several other low-frequency perturbations regrouped under the appellation "drift noise." Visualizing and quantifying those noises can be done through a series of calibration experiments involving immobile samples, such as magnetic beads stuck on a microscope coverslip. In order to quantify minute motion details about noise, we performed long, 10 h acquisitions (while typical measurements can be as short as 10–20 min).

We have implemented two schemes in order to correct the various sources of noise in our system. First, we applied a software-based baseline correction derived

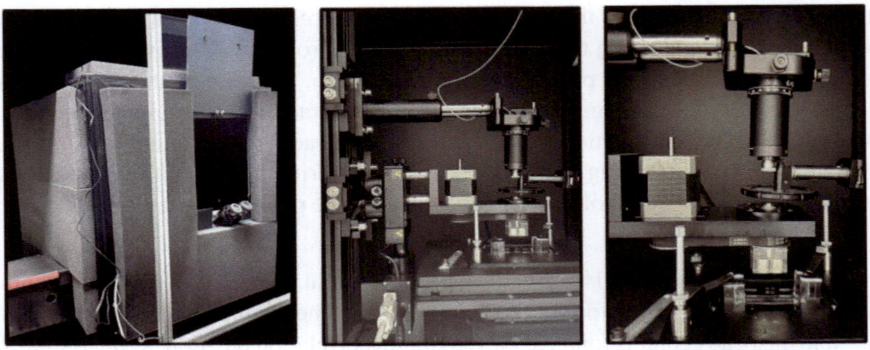

Fig. 8.3 (Left) Photographs of the isolation enclosure used for noise reduction built from Thorlabs rails, black cardboards, and locally supplied acoustic and thermal isolation foam panels (5 cm thick). An LED stripe is used inside the box for higher visibility during sample placement. (Center, Right) Close-up photographs of the microscopes showing the illumination system, the motorized permanent magnets, as well as the motorized sample stage

and adapted from high-precision metrology [24–26]. The correction basically relies on a fine measurement of the noise. Once characterized, the noise can be subtracted from the actual measurement. Second, as shown in Fig. 8.3, we built an enclosure around the setup for acoustic, seismic, thermal, and stray light isolation. Those noise correction schemes allow us to reach unprecedented resolution on the characterization of the protein extension during unfolding.

In order to quantify the resolution gain for an actual measurement, we found that using short DNA hairpin structures is ideal because they exhibit behaviors that are well characterized, in particular their length expansion when pulled. We then performed a series of measurements with and without the noise correction to finely quantify our resolution gains.

The combined effect of both the enclosure and the baseline correction is as follow: for shorter measurements (less than 20 min), the precision in the estimation of the extension of the protein during unfolding went from 0.9 nm to about 0.52 nm for measurements performed at 1.2 kHz and filtered with a moving averaging window 5 Hz. Furthermore, the added stability of the isolated setup, combined with a stable tracking algorithm and a more durable pulling construct allowed for longer measurements to be performed. We found that the resolution at which the extension is measured gets better as the square root of the acquisition time. At best, for 67 min measurements and above, we measure the extension of the protein with a 0.3 nm resolution.

We illustrate in the next section that the increase in resolution brings two major benefits: 1. It allows for the precise characterization of the extension of unfolded states as measured in a high-force unfolding experiment. 2. It allows for more precise measurements in time, which helps characterizing the statistical model of unfolding events.

8.2.5 Real Measurement

We performed a set of experiments on the Escherichia coli rhomboid protease GlpG to showcase, on real measurements, the implications of a gain in resolution. We repeatably induced a full unraveling of the protein to its unfolded state by alternating between high and low forces exerted by the magnets in a force-jump type experiment. Regions of low forces exhibited reversible local folding and unfolding, characterized by minute downward and upward displacements respectively. It was also verified, by comparing data with simulations, that the protein reached its native N state and fully extended to its expected final value when subjected to high forces. While the two most extreme states have sufficient data points to be localized at our best 0.3 nm resolution, we found that the two intermediate folded states could be localized at reduced resolutions 0.35 nm and 0.6 nm. Those measurements are only made possible thanks to the reduced Brownian motion that the tracking particles undergo at intermediate forces above 5pN. This showcases our gain in resolution compared to previous measurement where intermediate folding states were localized at about a 1nm resolution [20].

To go one step further in the analysis, we can add that the protein extension was estimated by using a statistical model called the *Hidden Markov model* (HMM) [27, 28]. While some recent simulation-based studies have shown that a finer description beyond those models were possible [29, 30], virtually all recent work rely on an approximation of said HMM. The transfer rates from a folded state to another are typically obtained by fitting the hitting time (the waiting time before jumping from one state to the next) distributions by a single exponential decay [20]. Under our newfound time resolution, we found that the hitting time distribution obtained from a series of unfolding experiments under high force was actually best fitted by a double exponential decay of characteristic times 0.17s and 1.8s, hinting at a two-phase type distribution Markov Process. This discovery was made possible only thanks to the increase in both resolution and stability of our setup. Van Der Heijden [31] has given a mean to fit such processes, and we identified that it closest resembled a specific type of Markov process in the form of the so-called two-phase Coxian distribution. Coxian processes display the unique property of allowing each state not only to move to the next but also to bypass every other intermediate and directly jump to the last state. This behavior is observed routinely during high-force unfolding experiments where proteins jump from a folded state to being fully unfolded, skipping intermediate states. We counted that the probability of such events was in excellent agreement with what was given through van der Heijden's model. Interestingly, low-force refolding experiments are not expected to show similar behavior. Indeed, when lower forces are exerted and the protein is let to re-fold, we typically observe non-monotonic pathways with several successive folding and unfolding events, with no direct jump to native states. Experimentally, we fitted the hitting time distribution in such experiments and found that it follows a single exponential decay, hinting that refolding experiments are more suited to the HMM analysis contrary to high force unfolding. Potentially, the force applied to the system has a non-negligible effect on the unfolding process.

While this two-phase type behavior has been seen in a particular setting [32], the underlying mechanisms of the process have not yet been described, to the best of our knowledge, and the current work allowed for such minute detail description thanks to our improved resolution, both spatial and in time.

In conclusion, magnetic tweezers are an extremely valuable tool to characterize in minute details the folding and unfolding pathways of proteins. Though in principle relatively simple, it requires skills that spans many scientific fields: optics, image processing, biochemistry, and fine data analysis. In a sense, this experiment is a display of the scientific method at work in its truest fashion: highly collaborative, iterative, based on data and careful interpretation, with direct implications in several aspects of our society.

References

1. Lauffenburger, D.A., Horwitz, A.F.: Cell migration: a physically integrated molecular process. Cell **84** (1996)
2. Petrie, R.J., Doyle, A.D., Yamada, K.M.: Random versus directionally persistent cell migration. Nature Rev. Molecul. Cell Biol. **10** (2009)
3. Bustamante, C., Smith, S.B., Liphardt, J., Smith, D.: Single-molecule studies of DNA mechanics. Current Opinion in Struct. Biol. **10** (2000)
4. Puntheeranurak, T., Neundlinger, I., Kinne, R.K.H., Hinterdorfer, P.: Single-molecule recognition force spectroscopy of transmembrane transporters on living cells. Nature Protocols **6** (2011)
5. Efremov, Y.M., Cartagena-Rivera, A.X., Athamneh, A.I.M., Suter, D.M., Raman, A.: Mapping heterogeneity of cellular mechanics by multi-harmonic atomic force microscopy. Nature Protocols **13** (2018)
6. Popa, I., Kosuri, P., Alegre-Cebollada, J., Garcia-Manyes, S., Fernandez, J.M.: Force dependency of biochemical reactions measured by single-molecule force-clamp spectroscopy. Nature Protocols **8** (2013)
7. Muller, D.J., Engel, A.: Atomic force microscopy and spectroscopy of native membrane proteins. Nature Protocols **2** (2007)
8. Yamashita, H., Taoka, A., Uchihashi, T., Asano, T., Ando, T., Fukumori, Y.: Single-molecule imaging on living bacterial cell surface by high-speed AFM. J. Molecul. Biol. **422** (2012)
9. Benaglia, S., Gisbert, V.G., Perrino, A.P., Amo, C.A., Garcia, R.: Fast and high-resolution mapping of elastic properties of biomolecules and polymers with bimodal AFM. Nature Protocols **13** (2018)
10. Pfreundschuh, M., Martinez-Martin, D., Mulvihill, E., Wegmann, S., Muller, D.J.: Multiparametric high-resolution imaging of native proteins by force-distance curve–based AFM. Nature Protocols **9** (2014)
11. Kong, L., Zhang, P., Wang, G., Yu, J., Setlow, P., Li, Y.Q.: Characterization of bacterial spore germination using phase-contrast and fluorescence microscopy, Raman spectroscopy and optical tweezers. Nature Protocols **6** (2011)
12. Seol, Y., Carpenter, A.E., Perkins, T.T.: Gold nanoparticles: enhanced optical trapping and sensitivity coupled with significant heating. Opt. Lett. **31** (2006)
13. Liphardt, J., Onoa, B., Smith, S.B., Tinoco Jr, I., Bustamante, C.: Reversible unfolding of single RNA molecules by mechanical force. Science **292** (2001)
14. Charvin, G., Bensimon, D., Croquette, V.: Single-molecule study of DNA unlinking by eukaryotic and prokaryotic type-II topoisomerases. Proc. National Acad. Sci. **100** (2003)

15. Del Rio, A., Perez-Jimenez, R., Liu, R., Roca-Cusachs, P., Fernandez, J.M., Sheetz, M.P.: Stretching single talin rod molecules activates vinculin binding. Science **323** (2009)
16. Min, D., Jefferson, R.E., Bowie, J.U., Yoon, T.Y.: Mapping the energy landscape for second-stage folding of a single membrane protein. Nature Chem. Biol. **11** (2015)
17. Shon, M.J., Kim, H., Yoon, T.Y.: Focused clamping of a single neuronal SNARE complex by complex in under high mechanical tension. Nature Commun. **9** (2018)
18. Kim, C., Shon, M.J., Kim, S.H., Eun, G.S., Ryu, J.K., Hyeon, C., Jahn, R., Yoon, T.Y.: Extreme parsimony in ATP consumption by 20S complexes in the global disassembly of single SNARE complexes. Nature Commun. **12** (2021)
19. Choi, H.K., Kang, H., Lee, C., Kim, H.G., Phillips, B.P., Park, S., Tumescheit, S.A., Kim, H., Lee, S.H., Roh, S.H., et al.: Evolutionary balance between foldability and functionality of a glucose transporter. Nature Chem. Biol. **18** (2022)
20. Choi, H.K., Min, D., Kang, H., Shon, M.J., Rah, S.H., Kim, H.C., Jeong, H., Choi, H.J., Bowie, J.U., Yoon, T.Y.: Watching helical membrane proteins fold reveals a common N-to-C-terminal folding pathway. Science **366** (2019)
21. Dulin, D., Barland, S., Hachair, X., Pedaci, F.: Efficient illumination for microsecond tracking microscopy. PLoS ONE **9** (2014)
22. Cheong, F.C., Krishnatreya, B.J., Grier, D.G.: Strategies for three-dimensional particle tracking with holographic video microscopy. Opt. Expr. **18** (2010)
23. Lucke, R.L.: Rayleigh-Sommerfeld diffraction versus Fresnel-Kirchhoff, Fourier propagation, and Poisson's spot. Defense Technical Inform. Center **30** (2004)
24. Graciani, G., Filoche, M., Amblard, F.: 3D stochastic interferometer detects picometer deformations and minute dielectric fluctuations of its optical volume. Commun. Phys. **5** (2022)
25. Graciani, G., King, J.T., Amblard, F.: Cavity-amplified scattering spectroscopy reveals the dynamics of proteins and nanoparticles in quasi-transparent and miniature samples. ACS Nano **16** (2022)
26. Graciani, G., Amblard, F.: Random dynamic interferometer: cavity amplified speckle spectroscopy using a highly symmetric coherent field created inside a closed Lambertian optical cavity. Appl. Opt. Metrol. III (2019)
27. Zhang, Y., Jiao, J., Rebane, A.A.: Hidden Markov modeling with detailed balance and its application to single protein folding. Biophys. J. **111** (2016)
28. McGibbon, R., Ramsundar, B., Sultan, M., Kiss, G., Pande, V.: Understanding protein dynamics with L1-regularized reversible hidden Markov models. In: International Conference on Machine Learning (2014)
29. Ayaz, C., Tepper, L., Brünig, F.N., Kappler, J., Daldrop, J.O., Netz, R.R.: Non-Markovian modeling of protein folding. Proc. National Acad. Sci. **118** (2021)
30. Suarez, E., Adelman, J.L., Zuckerman, D.M.: Accurate estimation of protein folding and unfolding times: beyond Markov state models. J. Chem. Theory Comput. **12** (2016)
31. Van Der Heijden, M.C.: On the three-moment approximation of a general distribution by a Coxian distribution. Probability in the Eng. Informat. Sci. **2** (1988)
32. Tapia-Rojo, R., Eckels, E.C., Fernández, J.M.: Ephemeral states in protein folding under force captured with a magnetic tweezers design. Proc. National Acad. Sci. **116** (2019)

Chapter 9
A Look Ahead

Abstract This chapter explores the transformative convergence of physics and machine learning in scientific inquiry and experimentation. Traditional experimental techniques are increasingly being augmented by machine learning models that incorporate physical laws directly into their architecture, a paradigm known as physics-informed deep learning. These models are not only more efficient and accurate but also more aligned with the fundamental truths of the natural world. Beyond analysis, machine learning now plays an active role in experimental design, optimizing instrument parameters to generate the most informative data, an emerging feedback loop between computation and empiricism. However, this technological power demands societal responsibility. The COVID-19 pandemic underscored that scientific advancement alone is insufficient without public trust, clear communication, and informed engagement. The chapter argues for a reinvention of science education and communication that emphasizes critical thinking, transparency, and the scientific process itself. Ultimately, building trust among scientists, policymakers, and the public is essential to harnessing the full potential of science for the collective good.

9.1 Conclusion

Throughout this book, we have explored a variety of experimental and theoretical techniques, each a product of careful design, empirical rigor, and often decades of incremental progress. But increasingly, these techniques are being reshaped, augmented, or even reimagined through the lens of machine learning.

One of the most promising direction is the incorporation of physical laws directly into these learning models. In the early days of artificial intelligence, a model would learn solely from the data it was given, without any awareness of the constraints and symmetries of the physical world. But the universe is not arbitrary. It follows patterns, symmetries, conservation principles, deep truths we have spent centuries uncovering. To ignore them is to train blindfolded. By embedding physical equations into learning architectures, we create models that are more resilient to noise, require less data to train, and are grounded in the very fabric of reality, less prone

G. Graciani, *Effective Communication and Single-Molecule Research*,
SpringerBriefs in Molecular Science,
https://doi.org/10.1007/978-3-032-10426-7_9

to producing hallucinations. This approach, often referred to as physics-informed or physics-inspired deep learning, is not a curiosity, it is rapidly becoming essential.

Take, for example, the remarkable domain of super-resolution microscopy. Optical limits as we have seen, once considered immutable due to diffraction, are now also being challenged by computational techniques involving deep neural networks. Here, the physics of light is not merely respected, it is actively embedded within the neural architecture, guiding it, constraining it, shaping it. The result is a form of "computational microscopy" that allows us to see deeper, clearer, and more precisely into the biological world than ever before. But the story does not end with analysis.

We are now indeed witnessing machine learning models that do not merely process data, but actively help scientists design experiments to exactly and efficiently generate the data they need. These systems help us tune the very parameters of the instruments: optical paths, illumination profiles, acquisition rates, so that the results will be most amenable to analysis. This is a paradigm shift. The experiment is no longer a fixed input into a passive algorithm. It is an active dialogue between instrument and processing. This new field of physics-inspired deep learning for instrumental design represents a new kind of virtuous feedback loop between the empirical and the algorithmic.

More traditionally, machine learning still helps us tremendously for the analysis of data. As we generate ever larger quantities of points through high-throughput experiments, the role of machine learning in parsing this vastness becomes even more pronounced. In the domain of drug discovery, for instance, the design of antibody-drug conjugates is a problem of such staggering combinatorial complexity that traditional chemistry alone cannot easily solve it. Yet machine learning thrives in such a landscape. Given large enough datasets, algorithms can identify patterns in how molecular structures interact with biological targets.

Such power comes with responsibility. Artificial intelligence requires training. It requires computational resources. It requires sustained investment. And this investment must come from societies, from governments, from institutions that understand the stakes. For that to happen, policymakers must grasp the transformative nature of this work. The general public must be engaged, not merely as bystanders, but as informed participants in the future that is unfolding, in a true democratic fashion.

The COVID-19 pandemic provided a sobering case study. The rapid development of vaccines and therapeutics was a great scientific achievement. And yet, misinformation, distrust, and fear undermined the acceptance of these life-saving interventions. It became painfully clear that scientific discovery, no matter how brilliant, is not enough on its own. Communication is essential. Trust is essential. Scientific culture is essential.

This is why science communication is not a luxury. It is a foundation. We must teach not only the findings of science, but the process of science: its iterative nature, its self-correcting mechanisms, and its openness to scrutiny. We must reveal that research is not a static archive of truths but a living method, a way of asking questions, testing ideas, and updating beliefs.

This is particularly vital when we think about the education of future scientists and citizens. It is not enough to fill minds with facts. We must cultivate autonomous, critical thinkers. People who are capable of navigating a world rich in information, disinformation, and complex technologies. People who can interrogate claims, evaluate sources, and apply reason. People who can learn how to learn, using the internet, AI, and other tools, not to outsource their thinking, but to enhance and empower it, bringing science into the heart of our democratic process.

In the final analysis, what is needed most is trust. Trust in science, but also trust in one another. Trust that through openness, transparency, and education, we can meet the profound challenges and opportunities of our time.

These are not challenges just for scientists, but for society as a whole.

9.2 TL;DR 2.0

Information age has now become the age of transactive memory. Scientific facts are stored away from us on the internet, and large language models can summon them at will. Learning them is now superfluous. But in a sea of potential misinformation, what needs to be taught instead is the scientific method, a systematic process for critical thinking. Beyond that, people's lives are ruled by increasingly complex technology. If we truly want citizens to take part in the scientific debate and express their informed opinion on these matters in a true democratic fashion, we need them to be educated on the subject. The remaining question is how? By re-imagining how science is taught to students. Through the personal lens of the scientists who made it: their motivations, their experiments, their failures, and their successes. Through a direct access to science literature and to the voice of the scientists who made it.

MIX
Papier aus verantwortungsvollen Quellen
Paper from responsible sources
FSC® C105338

If you have any concerns about our products,
you can contact us on
ProductSafety@springernature.com

In case Publisher is established outside the EU,
the EU authorized representative is:
Springer Nature Customer Service Center GmbH
Europaplatz 3, 69115 Heidelberg, Germany

Printed by Libri Plureos GmbH
in Hamburg, Germany